NETSPACES

Netspaces

Space and Place in a Networked World

Katharine S. Willis
Plymouth University, UK

Routledge
Taylor & Francis Group

LONDON AND NEW YORK

First published 2016 by Ashgate Publishing

2 Park Square, Milton Park, Abingdon, Oxfordshire OX14 4RN
711 Third Avenue, New York, NY 10017

Routledge is an imprint of the Taylor & Francis Group, an informa business

First issued in paperback 2017

British Library Cataloguing in Publication Data
A catalogue record for this book is available from the British Library.

Library of Congress Cataloging-in-Publication Data
Willis, Katharine S.
    Netspaces : space and place in a networked world / by Katharine S. Willis.
        pages cm
    Includes bibliographical references and index.
    ISBN 978-1-4724-3862-1 (hardback : alk. paper)
        1. Mobile computing. 2. Internet--Social aspects. 3. Cyberspace. I. Title.

    QA76.59.W555 2015
    302.23'1--dc23

                                                                        2015018235

ISBN 978-1-4724-3862-1 (hbk)
ISBN 978-1-138-57339-0 (pbk)

Contents

List of Figures

Preface

This book emerged out of an architect's curiosity about technology, and what it might mean or how it might change the experience and design of the built environment. The pursuit of this curiosity has led me through many different experiences, encounters and places. I have travelled hundreds of kilometres to visit Data Centres, spent many hours mapping data on location-based media such as Foursquare, long days walking the streets of various cities logging Wi-Fi nodes, spent weeks setting up interviews with the organisers of a flash mob and most recently visited a Smart Home showcase. Yet, an underlying interest remains with GPS technology, and how it affects the way we experience, understand and remember the city. I bought my first GPS receiver, a Garmin Etrex, in 2000 and became increasingly fascinated with the idea of the connection between my body, the information on the GPS receiver screen, the space around me and the constellation of GPS satellites orbiting 40,000 kilometres above my head. Despite the fact that many ways in which we talk about technology assume accuracy, and an exact copy of the 'real' world, during my investigations I found so many examples of ambiguities, and of how GPS technology showed me other experiences of places. I set about testing out the incongruities between the spatial environment and the GPS space. I came across numerous articles in the national and local press that documented people using satnav's in cars to make extraordinary decisions; a woman who drove for nine hundred miles from her home in Belgium to Croatia, when she had wanted to go to Brussels, because her satnav 'told her to', people who drove off cliffs and the wrong way down one-way streets. It seemed that the space of GPS was working on a completely different framework to the one of the 'real' world. In the early part of the noughties the use of satnav's, google maps and other 'location-based services' was just taking off. I began to wonder if people who used these GPS based interfaces and social media were somehow experiencing and learning about the space differently. I decided to see if I could study how these changes were taking place and the effects this was having on urban space.

Sadly, I left my yellow Etrex on a bus in Italy in 2010. But the loss was countered by the fact that really I no longer needed it. It had become redundant. Now, my phone has built-in GPS, internet and social media to know where 'I am'. In fact the digital spaces of the city are no longer separate from me, but embedded in almost every way I experience and inhabit the city. These are what I use to make sense of the city, and the city I experience is both technological and material.

Acknowledgements

As with any publication, this book has been a journey in itself; and many people have helped along the way. The main reason this book turned from an idea into a reality was an Akademie Schloss Solitude fellowship. This gave me both a place to work on the book and most importantly a space to think. Thanks to Jean-Baptise Joly for the great discussions as well as Marianne Roth and Silke Pflüger who organised everything. The underlying ideas for the text emerged from my doctoral research. In 2006, newly arrived in Germany, I pitched up at University of Bremen and somehow persuaded Christian Freksa of the SFB/TR8 programme to take me on as a doctoral student, which was the start of the whole journey. Over the next few years the core ideas within the book emerged from my research as part of the Marie Curie Mediacity project with support from Frank Eckardt and Jens Geelhaar at the Bauhaus University of Weimar. Thanks also to my current colleagues at Plymouth University who have supported me through time and feedback, Alessandro Aurigi in particular, as well as Malcolm Miles and Bob Brown. The thinking in the book has also been shaped by the ideas of others, but especially Miriam Struppek whose work on the topic of urban screens was inspiring. Thanks also to Sophie Harbour for her work on proof reading, and more generally for cheering me up in the last stages of the work on the draft. Also, thanks to Valerie Rose for her editorial support and of course for commissioning the text in the first place. Finally thank you to my family; Bernd, my husband and my sons Lukas and Robert, for being there when I wasn't.

Introduction

The focus of this book is on understanding and explaining the way that our increasingly networked world impacts on how we experience and inhabit urban space. It reflects on the nature of the spatial effects of the networked and mediated world; from mobile phones and satnavs to data centres and Wi-Fi nodes and discusses how these change the very nature of urban space. It proposes that netspaces are the spaces that emerge at the interchange between the built world and the space of the network. It aims to add to our understanding of the fundamental changes occurring in built space due to the ubiquity of networks and media.

We tend to take for granted that the built space around us can be experienced through the senses; it is tangible and visible. We can feel the roughness of a brick, hear our footsteps on a marble floor and know that a door handle will open a door. It is also somehow 'permanent'; that is buildings tend to stay in place for a period of time so as to be thought of as just 'being there'. It is for this reason that we notice when a building is knocked down, or a new one is constructed. Moreover our experience of the built environment is shaped by the idea that what we see is a way to make sense of the building or city or landscape. What we can see with our eyes is 'what it is', and we use these features to orientate ourselves in our everyday lives.

The built world is understood a great deal for how it looks, and how we make sense of it visually. Thus we expect a hospital to look hospital-like, a station to look like a place for trains to arrive and depart and a park to be landscaped so as to be pleasant to spend time in. Of course this is all linked tightly to what actually happens in these spaces, but still the image or visual impression of the built world helps us to make sense of the large number of buildings we encounter in a place like a city without having to work out what happens in each building anew each time we see one. In fact the process of globalisation has exacerbated this condition, with places literally copied and located in world cities so that we know what a certain coffee shop will look like whether it is in Mumbai or Rio or Copenhagen. In fact when you are inside, it is sometimes difficult to remember which city you are actually in. But our increasing reliance on the internet, mobile phones, satnavs and other networked technologies changes our relationship with the material, built and tangible world. For example, say you are in one of these coffee shops, but not so much to drink coffee, rather to log in to the Wi-Fi connection. Are you in a café or are you actually in a network space? When you're online, you are less grounded in the material place that you are in. You're sitting in the coffee shop, but your mind is focused on a friend who is in another city, probably a different time zone. Maybe they're sitting in the same chain of coffee shop, but in their own city.

These patterns of connection create traces, ephemeral patterns of connections and situations we engage with. These networks help us to structure our day; we agree where to meet someone later, find out when a film is showing or book a place to eat for lunch. None of this is done by getting up and actually going to the place or person in question. We do not have to be there or to have spoken to someone to make an impact on how we inhabit the space. Instead we leave a trace of our presence in our interactions with these situations: an online payment, a text message, a phone call. In all of these interactions there is a technological link between the sender and the receiver; a mobile phone mast negotiates a packet of data that is a text message and routes if through the cell network to the receiver's mast location and then their phone. But this network connection has very little visual quality. Where is all the data from mobile phones, radio signals and Wi-Fi? It passes through the air we breathe, but we never see it. We only become aware of it when it materialises on a device or a screen.

So in the café, what matters is whether the connection is good, whether there is a power socket nearby and whether it is too noisy or uncomfortable to concentrate on the conversation happening online. In this way there is no distinction between the place of the Internet connection and the chair, table or the physical environment of the coffee shop. But the key thing is that it really doesn't matter that much what the place looks like. We're not there to be in the place so much as to 'use it'. Things have to work – like the table not falling over and there being some light. But because the attention is on the actual space of a conversation or connection the quality of the space is somehow less important.

CHALLENGES

There is a growing awareness in the design and architecture field of the need to address the way that technology affects the design of urban space. The recent emergence of 'smart cities' as well as the growth in 'Big Data' and the Internet of Things (IoT), have highlighted the connection between ICTs and the city. In this book I argue that networked technologies in the city do not result in non-places, and to clearly explain the subtle and multi-faceted ways in which networked and digital spaces can be meaningful and high-quality places. How are architects and urban designers dealing with these mediated networks? The simple answer is hardly at all. They don't work on the same frameworks that designers use to develop a programme for a space. At best they are seen as necessary but non-negotiable infrastructures. So mobile phone masts are camouflaged, data centres are literally the architectural realisation of a black box and Radio-Frequency

Identification (RFID) tags are sealed inside objects. Visualising and mapping traces of these interactions has been the only way to date that has sought to make sense of theses networks within the city, but this is purely trying to map one framework onto the structure of another, wherein lies a the problem that they are inherently different. This book aims to provide an understanding of the significant changes in how people inhabit and experience urban space and the consequent implications for the design of cities.

STRUCTURE OF THE BOOK

The structure of the book is organised around six core chapters and a final chapter that draws the findings of each chapter together to consider how they might address some overarching societal issues. Each of the chapters addresses a different dimension of the changing spatial word, and how this affected by networked technologies. The chapters themselves are each organised around a series of sections; firstly key ideas are introduced and then a historical overview is given to situate the changes within a context. This is followed by a more detailed investigation of the changing conditions around the topic, both theoretically and empirically. The final section of each chapter discusses a case study that highlights the changing nature of urban space and the role that technology plays in this change. These case studies focus both on particular characteristics or typologies of urban space, and also on a specific set of technologies, infrastructures or platforms. The aim of this approach is to provide a comprehensive investigation of the issues around a particular topic, and to move between theoretical enquiry, historical study and empirical application in a way that addresses and reflects on as many dimensions of the condition as possible. The chapters link to each other in terms of content, but do not necessarily require the reader to have read one in order to understand another. This means they can be read separately or in a different sequential order. The book concludes with a seventh, summary chapter that reflects on the findings of individual chapters and seeks to synthesise these into a key discussion. There is also some discussion surrounding the implications of these changes on future developments within the fields of architecture and urban design.

Infrastructure

The first chapter works with the concept of networked infrastructure. It introduces the concepts of meshworks and assemblages to offer an alternative approach to the understanding of networked infrastructures. It highlights how Lefebvre's concept of meshworks allow for a more subtle, human-centred links and connections that integrate what Castells refers to as a 'space of flows' with the 'space of places' (1996). This is contextualised through a reflection on how technical infrastructures have historically impacted on urban form, such as how the wired frameworks of the telegraph and telephone shifted the organisation of urban spaces from centralised to networked and decentralised frameworks. The more recent dominance of digital infrastructure in the urban condition is highlighted to demonstrate the consequent emergence of three problematic conditions. Firstly, the lack of visual and material presence of these infrastructures which has implications for the visual organisation of cities; secondly, the black boxing, and thus de-socialising of these infrastructures; and thirdly the increasing shift of infrastructures away from integration with existing built structures and in to the air and hidden beneath our feet. These issues are explored through a case study of 'the cloud', which is described in terms of the growth of a globalised network of 'cloud factories' or data centres. The chapter documents how data centres are increasingly becoming the global scale, built black boxes of our networks, and how the material infrastructure of the network; Wi-Fi nodes, undersea cables, data centres and control rooms are increasingly becoming part of our built infrastructure. It also highlights a range of emerging issues, such as the merging of energy and communication infrastructures, coupled with the exponential growth in energy requirements to 'power the internet'. It concludes by reflecting briefly on the implications of the vast, invisible physical infrastructure of our digital networks, and how this interplays with our current approach to the design of urban space that privileges visual and legible urban frameworks.

Places

The meshworks framework is extended in Chapter 2 to a discussion around the nature of place, including an outline of how networks create a new sense of place. This chapter works with an approach to space and place in terms of an approach that sees space as socially constructed (Lefebvre, 1991; Massey, 2005), and argues that places are increasingly becoming contingent on interactions occurring within technological meshworks (Gordon and de Souza e Silva, 2011). Networked spaces, far from being anytime, anywhere, are emplaced and meaningful, but they do operate on different frameworks to a phenomenological reading of place such as that of Tuan (1977) and Relph (1984), and even to some extent Castells (2004). It positions an approach to place that counters the argument that places are turned into non-places by the interactions and flows of networks, and draws on the work of a range of authors who show that networked places are in fact rich, social places working within a physical, spatial framework. Through a short historical review on the way that technology has changed place-based relations over the last fifty years, the chapter outlines how it possible to see how factors such as increased mobility and the shift of computer-based communication away from fixed computers to ubiquitous, social technologies has resulted in a merging of physical places and online social spaces. It is argued that this results in a shift away from fixed and discrete places to an increasing importance of inbetween places (Ito and Okabe, 2005), combined with the ability of people to move their attention between online and offline spaces almost seamlessly. This results in a heightened sense of the here and now, a phenomena that is characterised as 'selfie architecture'. The consequent implications for Lynch's concept of urban legibility (Lynch, 1967) are discussed, reflecting on the fact that these have relied upon a visual reading of place, and that places now become legible and salient in the urban environment through social interactions rather than those of the eye and the physical senses. The case study of the use of Foursquare, a location-based social network is used to illustrate the changing nature of place, where places are socially constructed through the practice of naming and 'checking-in' at 'venues'. Foursquare venues are valued by the number of 'check-ins' and are typically inbetween spaces; that is sites of transition and temporal occupation. Airports, railway stations, cafés, restaurants and sports centres are the places that become most valued in the meshwork, and they are also characterised by their sociality; they are places where people converge and then disperse; brought into being for the time in which the networked links connect. The chapter concludes that the places of the network are increasingly experience as inbetween spaces, that are the everyday spaces of transit and social encounters both on and offline. Rather than placeless they open up opportunities for new spatial typologies and experiences that offer a more situated and socially constructed sense of place.

Boundaries

The next chapter explores further the spatial qualities of network frameworks, and outlines how traditional ideas of containment and proximity are shifted. The chapter outlines how spatial concepts such as threshold and edge, inside and outside, public and private are underpinned by relationships of physical proximity (Hall, 1966), and connection, so that according to Simmel 'we are at any moment those who separate the connected or connect the separate' (1994 (1909), p. 171). The chapter argues that networked technologies change the characteristics of these relationships, so that boundaries are defined by the degree of linkage in or access to the network, and less by the physical structures of connection and separation created by walls and doors. For example, in netspace it is possible to enter and leave a building without having moved, and also to be in two spaces at once. The chapter investigates the historical context for these changes, and documents how the emergence of wireless and networked technologies have reconfigured spatial rules such as urban zoning patterns, and also challenged social rules such as what constitutes public and private spaces. The changing nature of boundaries is highlighted in how digital interactions are resulting in different models of private and public, with a 'public-by-default' (boyd, 2014) replacing the traditional model of choosing to be 'in public'. Similarly, the uses of technology and social networks create new types of shared spaces underpinned by a sharing of resources through peer-to-peer networks that are responsive and networked. Whilst almost constant connectivity may one of the inevitable outcomes of a networked society, this also throws up significant challenges for those who become excluded, or are not given access to the digital world. It is shown that 'digital divides' is a significant and real issue and that whilst some boundaries may break down, conversely new boundaries of exclusion emerge. Through a case study of the use of public Wi-Fi network, the chapter tests the way that these networks affect patterns of access, and how public and private territories and also inclusion or exclusion become defined and experienced. The chapter also explore how patterns of media use enables people to construct personal territories, and enables organisations to define group territories that operate much as a building might be bounded by walls and a roof.

Publics

This study of shifting boundaries is extended in Chapter 4 to consider how networked technologies reconfigure existing definitions of what constitutes public space. Drawing on readings by Arendt (1999), Sennett (1977) and Habermas (1999) on the public realm and public space, it looks at the emergence of what has been termed 'networked publics' (Varnelis, 2012). It considers whether public space is the only stage where 'publicness' can be enacted or whether networked spaces also create the conditions for people to experience the public realm. The chapter then explores in more detail ideas of agency and performance in the built environment, and questions how network technologies enable urban space to become a site for encounter and participation, in opposition to a long line of urban theory that sees urban public space as increasingly remote and dehumanising. Working from the historical context of television and broadcast media, concepts of actor and audience, as well as frameworks for how people can participate in the public realm when it is experienced through digital technologies are examined. This chapter draws on case studies of urban screens and mobile media to show how these types of media work with three core readings of public space; ones that enable exposure to encounters with strangers, those that create spaces for intervention and interference and finally playful sites for the public construction of social norms. The chapter documents how urban screens re-introduce a social stage to the city, merging playful shared experiences with everyday activities and creating conditions for 'shared encounters' (Willis et al., 2008). The combination of online publics and urban screens is discussed in relation to the political movements, such as Occupy, where the link between a focal urban public space and online social networks, is shown to be critical for the organisation and enactment of public action. The chapter concludes that forms of publicness are emerging that work across physical public space and online networks, that requires new thinking about the built environment of public space in terms of its capacity to provide a 'stage' for the enactment of publicness.

Time

The temporal nature of the city is discussed in Chapter 5, and the changing rhythms of urban life discussed. We often forget that the patterns of day and night, of commutes to and from work and school and of walking characterise the urban experience. Working with Lefebvrian concepts of rhythm (Lefebvre, 2004) and Lynch's reading of time and place (Lynch, 1988) this chapter looks at the way that networks introduce alternative rhythms and consequently different spatial experiences. It discusses how digital rhythms cause activities to be highly synchronised in real-time rather than through being co-ordinated, planned and linear. It also considers how this opens up new perspectives on spatial memory and the implications of a our use of social media where nothings is ever forgotten or degrades over time. This is contextualised with a review of the changes in the last century with how everyday, socially constructed time was replaced by an abstract, global clock time with the advent of the industrial society. The link with the emergence of a highly mobile, commuter society is also discussed, and ever increasing growth in air travel and time spent on the move and in transit. This is explored through a case study of emerging digital rhythms and a focus on the condition of real-time cities (Townsend, 2000) and looks at how the patterns and traces that our movements leave behind show that our cities are increasingly choreographed by the network.

The emergence of practices such as swarming, or 'thumb tribes' is described as when groups co-ordinate activities in real time and how this creates new patterns of presence in the city. The implications of this highly synchronised and real time society is also discussed in the light of the fact that digital rhythms appear to be commodifying time as space; it is speeding up and filling up so that journeys are no longer down time but work time and transit spaces are now filled with people going online. The case study looks at the phenomena of 'real time' cities through a study of media use in transit spaces such as airports, stations and roads. This explores how airports are an example of a particular type of space that have started to become organised around time rather than space, and the consequent rhythms that start to emerge. In a study of the implications of real time coordination made possible by networked devices the chapter documents the emergence of the 'sharing economy' of hyper co-ordinated taxi rides and room-sharing apps and social network platforms. Finally the proliferation of pop-up, ad-hoc and temporal events is explored to understand how they re-introduce new temporal rhythms into 'timeless' spaces of airports and other transit zones. This shift into highly coordinated and managed rhythms has significant implications for how we occupy spaces, since their temporal quality is not just a characteristic of the space but a defining factor in how it is experienced.

Things

The final chapter looks beyond the typical domain of architecture and urban space to consider the networked world of objects and things that is referred to as IoT or Big Data. In this chapter there is a consideration of the nature of materiality when data and information are increasingly seen as resources and actors in our everyday world. This is explored through the lens of considering how we use the 'stuff' of the material world around us to construct a sense of identity and belonging. The emergence of a highly connected environment of sensors and connectivity is discussed in relation to the extent to which we still have agency in such spaces, and also the problems that coded and password protected space pose for inclusion and exclusion. The case study section here draws on a study of 'smart home' examples to test out these ideas. It looks at different models for how we construct a sense of identity in a networked home, as well as the nature of agency or control in our interactions with a 'sentient' domestic environment. The final consideration in this chapter is given to the hidden side of the digital; dirt, mess and decay and considers whether a smart home can become dirty.

The last chapter of the book moves beyond a general summary and discusses instead a range of future design challenges for urban space focusing on whether some of the changes documented in the book might address societal issues. Overall the book seeks to explore and document how our understanding of the city and how we make our way through it is changing as a result of networked technologies and infrastructures. The physical bricks and streets and buildings may not be crumbling or shifting like quicksand, but we are constructing different relationships with them through the networks and devices that we interact with in our everyday lives. The places and spaces of the city are shifting into a new merged space of places and space of the network; netspaces.

1

Infrastructures

1.1 NETWORKED INFRASTRUCTURES

22@

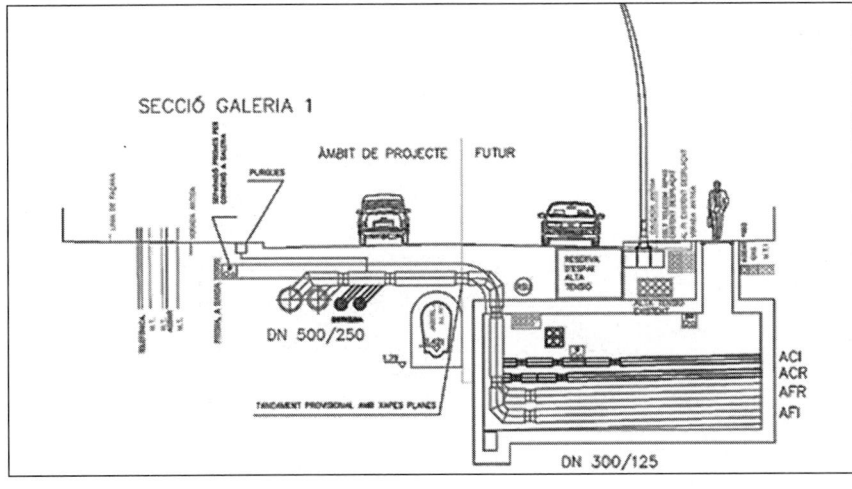

1.1 Cross section through Almogavers Street, 22@, Barcelona showing underground infrastructure 'galleries' (source: Barcelona City Council).

There were two rather unsettling things about my walk along Carrer de Tanger in the Poblenau district of Barcelona in autumn of 2013. Firstly, it was the fact the cills of the ground floor windows of the four storey historical brick building housing part of the UPF Communication Campus Poblenou were level with the street. So as you walked along and peered through the beautiful brick arched windows into the building inside you noticed that the floor level was about half a storey below. The second thing that made me curious was a large, anonymous box of a building, located directly opposite the busy University campus hub. It was obviously new, but the three storey high copper clad exterior appeared to have no entrance and no windows. The area I was exploring was to the east of central Barcelona; the neighbourhood of Poblenau. But since 2001, the district has been rebranded as 22@, a project that has converted 200 hectares of industrial land into a district for a 'knowledge economy' (Ajuntament de Barcelona, n.n.), and is often described as an example of a 'smart city' development. The reason for the strange sinking of the old brick-built factory buildings, now converted into high-tech media labs for the University? Part of the development involved the introduction of a new €180 million urban infrastructure which used the existing street system as a conduit for a converged power, heating, telecoms and waste disposal network. The existing street surface has literally been raised by two metres to accommodate 4.240 m of channels or galleries running beneath the road structure, comprising over 200 km of electricity cable, 700 km of telecommunications cable and 6.830 m gas pipeline (Ajuntament de Barcelona) (see Figure 1.1). According to the city 'while each of the industry clusters are segregated into distinct areas containing residential areas and amenities, they are unified by centralised heating and air-conditioning, electricity distribution, waste disposal, telecommunications infrastructure', and smart traffic management systems' (Leon, 2008, p. 238). So in 22@ there will be no sign of the ubiquitous men in yellow high-vis vests and a pneumatic drill, digging up the road yet again for a telecoms or power provider. Changing or updating or servicing the infrastructure of the neighbourhood is instead carried out by a workman walking into a gallery, so that new businesses can simply plugin to the system and capacity is controlled from a remote location, as people continue to walk the pavements above.

Which comes to the rather lonely looking copper clad box sitting in the middle of the street opposite? This turns out to be the power hub of the district; the Tanger power plant, providing 26 MW for heating and 27 MW for cooling the seventy eight buildings in the 22@ district and linked up to a system spreading through the under street infrastructure. Powered by natural gas, and designed with the aim of reducing CO2 emissions by twenty two per cent the plant also guarantees a robust energy supply under peak conditions. And in another curious twist it turns out that the combustion gas from the boilers is exhausted by the historical chimney of the textile factory Can l'Aranyó, originally built in 1872. The massive brick tower, made redundant as the textile industry in Poblenau was superseded by Far East factories, has been brought back to life as a chimney which pumps out the gas from the plant powering the new economy; ICTs and education.

Interestingly not only did the design of the 22@ infrastructure create a new way to organise and connect the space of flows to the space of places, it also used this approach to create 114,000 m^2 of new green space and 20,515 m bike lanes by opening up the existing dense urban grid into urban blocks and widening the sidewalks to seven meters. As a result, the fine grain of the urban street network has been replaced by larger, denser blocks that alter the sense of urban scale. The infrastructure of 22@ is in fact far from invisible or immaterial. The buildings need to be connected by vast swathes of power and fibre optic cables and installed in a way that allows for future change in occupancy and use of the buildings of the district. The power, telecommunications and transport infrastructure of a connected urban environment, such as 22@, create material changes in the structure and form of the urban fabric. If we remember that Le Corbusier's Ville Radieuse created a vision of streets in the sky and fast moving traffic at ground level, in the connected city the traffic of data and energy is buried beneath the surface and the streets become re-discovered at ground level.

Invisible cities

What lies beneath the city is often overlooked. The city is framed by flows and channels; those of people, energy, money, vehicles and data. Urban places are linked by 'movement channels of various kinds; doorways, street grids, transport networks' (Mitchell, 1995, p. 117). Indeed the flow of information in, through and out of a city is a fundamental characteristic of urban life. These flows are enabled by infrastructure, which is usually understood as one of the key building blocks of urban life and structure. Indeed cities are the densest expressions of infrastructure, or more accurately a set of infrastructures, that sometimes work well, but are sometimes chaotic or ineffective. Urban infrastructure consists of various structures; buildings, pipes, roads, rail, bridges, tunnels and wires brought together in a connected framework. But this framework also has rules; 'the software for the physical infrastructure, all the formal and informal rules for the operation of the systems' (Herman and Ausubel, 1988, p. 1). The hardware and the software of infrastructure often remains beyond reach of the citizen in their everyday life; it tends to be dirty, technically complex, and concealed. In the midst of the last century Lewis Mumford was one of the first urban theorists to highlight the importance of this 'invisible city' (1961, pp. 563–567). Mumford described how 'the new world in which we have begun to live is not merely open on the surface but also open internally … below the threshold of ordinary observation' (1961, p. 563). The idea of the city as container was being supplanted by 'new functions' brought into being by what Mumford called 'the functional grid; the framework of the invisible city' which rather than reintegrating the essential components of the city instead has 'tended to efface them' (1961, pp. 564–565). The prime players in this transformation were the power and communications systems. According to Moss and Townsend: 'the physical infrastructure that helped to shape earlier urban forms – the sea-port, the railroad, and the highway – is being superseded by a new network of optical fibres, Cisco routers, cellular antennas, and mobile telephones' (1999, p. 46). Communication infrastructures in the twenty first century have become a critical part of how a city functions, but as infrastructure they also represent a fundamental change in how a city is organised and controlled. Communication is now being treated as part of the social and technological infrastructure of city life.

Networks and the city

The key problem to be addressed in this chapter is how the shift towards information and communications technologies (ICTs) changes the old idea of the integrated, centralised city that has an identifiable boundary and is separated from other cities by distance. Whilst simultaneously the flow of media exchanges happens without much physical evidence on the surface of the city, Graham and Marvin have highlighted how 'rather than ending the domination of cities, these networks actually tend to erupt within the spatial order of the old city' (1996, p. 71). Communications technologies, from the masts connecting mobile phone networks to the fibrous cables of the Internet, whilst crucial in supporting the mobility and flux, are also fixed networks that must be embedded in space.

Space of flows

In his classic essay on The Network Society, Manuel Castell introduced the idea of the 'space of flows' (Castells, 1996). He set out a strange new world, where he countered that the 'the space of flows can be abstract in social, cultural, and historical terms … places are … condensations of human history, culture and matter'. They are no longer 'places', where 'place' is defined as 'a locale whose form and meaning are self-contained within the boundaries of physical contiguity' (Castells, 1996, p. 423). So Castells sets up a scenario that challenges the long-held privileged status of Cartesian geometry, the map, and the matrix or grid. Whether New York's famous grid street system, or the centralised street pattern of European market towns, these were being replaced by a global network infrastructure, rendering the space of places irrelevant. Instead, 'infrastructural links and connectors, as well as information exchanges and thresholds, become the dominant metaphors to examine the boundless extension of the regional city' (Boyer, 1999, p. 75). According to Graham and Marvin, the rise of globalisation 'undermines the notion of infrastructure networks as binding and connecting territorially cohesive urban space's … it forces us to think about how space and scale are being refashioned in new ways that we can literally see crystallising before us in the changing configurations of infrastructure networks and the landscapes of urban spaces all around us' (2001, p. 16). Castells concept of the space of flows essentially denies the spatial experience, and concludes that as we occupy global network infrastructures we become simply mediators of information pulsing through the network. Graham highlights the embedded-ness of these space of flows in the space of places so that 'the urban world connected by Gate's technologies string out on the wire is not disconnected, abstract, inhuman; it is bound in the places and times of actual lives, into human existences that are as connected, sensuous and personal as they ever have been' (Cosgrove, 1996, p. 1495). This became visible in the Poblenau district of Barcelona, now rebranded as 22@, with attempts to rename many of the existing urban spaces to reflect the new city structure. Some street plans showed roads that had, in fact, been lost because of the 'blocks' created by the densification of urban structure around various nodes of activity. Similarly the flow of people, previously to the textile factories, has shifted to new entrances and times as the students, IT workers and tourists of the knowledge economy flow in and out of the gateways. Meanwhile, an infrastructural building such as the Tanger power centre, which is a clear physical building within the street, is treated as an invisible entity. These are still meaningful places to both those who worked there, who visited them and those that linked in from global locations. But the flow of cables beneath the flow of people on the street above connected these places to a set of distant places that had as much meaning as those of the immediate urban fabric.

Aims

This chapter aims to explore the many different ways in which digital networks affect the structure of the city. It starts by opening up the concept of the network society, introduced by Castells (Castells, 1996), and discussing this in relation to a range of other readings and also introduces the terms 'meshworks' and 'assemblages' to consider more multi-layered and social perspectives on network infrastructures. An alternative reading of infrastructure through the concepts of assemblages and meshworks sees network infrastructure, not as invisible and placeless, but as embedded and interwoven with patterns of movement, energy and waste. It is important to remember that current digital infrastructures have not emerged from nothing, but are also part of a lineage that goes back to telegraph and telephone infrastructures. These wired networks have been shown to have had an effect on the spatial organisation of the city, shifting the organisation of urban spaces from centralised to networked and decentralised frameworks. More recent developments in the expansion of our digital infrastructure have caused a series of problematic conditions; firstly the invisibility of these infrastructures which has implications for the visual structure of cities, secondly, the black boxing, and thus de-socialising of these infrastructures, and thirdly the increasing shift of infrastructures into the sky and air and away from integration with existing built structures. These conditions are explored in more detail in the final section of the chapter where I look at the physical infrastructure of the Internet; data centers; which are where the data of the network is transferred and stored. The study of data centers shows that networked infrastructures are not only invisible and black boxed, but increasingly highly reliant on energy infrastructures. As such they pose important challenges for how we consider networked infrastructure as part of the spatial organisation of the city, and also open up important questions about the energy resilience of our increasing reliance on these infrastructures for many of our internet-driven everyday activities and urban processes.

1.2   MESHWORKS AND ASSEMBLAGES

Gaps in the network

Network theory, which addresses the widespread dispersal of digital information and communication technologies, of which mobile communication and the mobile Internet are the latest incarnation, has been considered by scholars as one of the central components of networked urbanism today (Castells, 1996); (Crang and Graham, 2007); (Graham and Marvin, 1996); (Graham and Marvin, 2001); (Mitchell, 1995) and (Shepard, 2011). Network theory of the nineties provided a valuable path through which to start to navigate and make sense of the interlinked global and local changes occurring in cities. However it also oversimplifies a much more complex set of inter-related flows and channels that are not limited to data, money, energy, water and vehicles, but also by their very nature include people, as well as the by-products of global connectivity such as rubbish and waste. Latour and Hermant point out that by paying attention to the material flows in and through infrastructures, you can appreciate more clearly how a city works. They argue that if you 'study a city and neglect its sewers and power supplies (as many have), and you miss essential aspects of distributional justice and planning power' (Latour and Hermant, 2004). Star argues that we can address this challenge by looking at the manifestation of networks at an everyday level, and through the way that they are standardised and categorised within existing frameworks. She contends that 'Perhaps if we stopped thinking of computers as information highways and began to think of them more modestly as symbolic sewers this realm would open up a bit' (Star, 1999, p. 379). To do this she argues that we need to look at it within its cultural and social context, just as 'the cook considers the water system as working infrastructure integral to making dinner. For the city planner, or the plumber, it is a variable in a complex planning process or a target for repair: "Analytically, infrastructure appears only as a relational property, not as a thing stripped of use"' (Star and Ruhleder, 1996, p. 113; Star, 1999, p. 380). This underlies a fundamental problem with the way infrastructure is applied; 'infrastructure networks are thus widely assumed to be integrators of urban spaces' (Graham and Marvin, 2001, p. 8). But if we start to address them as complex, messy, incomplete and 'knotted' rather than sleek, impenetrable fibres, cables and pipes then this opens up a more authentic reading of the structures that weave through our urban life.

Through Actor Network Theory (ANT) Latour, drawing on original work by Law, introduces a more stratified understanding of this condition, by pointing out that 'to say that something is a network is about as appealing as to say that someone will, from now on, eat only peas and green beans, or that you are condemned to reside in airport corridors: great for traveling, commuting, and connecting, but not to live. Visually, there is something deeply wrong in the way we represent networks, since we are never able to use them to draw enclosed and habitable spaces and envelopes' (Latour, 2011, p. 800). Thus ANT sees networks not as obliterating spatial relations and processes, but reconfiguring them 'involves relational assemblies linking technological networks, space and place, and the space and place-based users (and nonusers) of such networks (1993: 120). But one of the challenges of ANT is that to Latour, such technological networks are comprised of specific places 'aligned by a series of branchings that cross other places and require other branchings in order to spread' (Latour, 1993, p. 120). The spaces between connections are characterised as being strangely immaterial. Latour points out that 'they are connected lines, not surfaces. They are by no means comprehensive, global or systematic, even though they embrace surfaces without covering them, and extend a very long way' (Latour, 1993, p. 118) and Mackenzie highlights the problem that 'network theorising can de-animate relations' (2010, p. 11).

To address this, network infrastructure could be considered as something closer to what Lefebvre terms 'meshworks', the way in which the movements and rhythms of human and non-human activity are registered in lived space (1991, p. 117). The history of space thus begins with the spatio-temporal rhythms of nature as transformed by a social practice, imposing the 'meshwork' of mental and social activity upon nature's space (Lefebvre, 1991, p. 117). Lefebvre highlights how natural space changes as it is 'traversed now by pathways and patterned by networks ... one might say that practical activity writes upon nature, albeit in a scrawling hand, and that this writing implies a particular representations of space' (1991, p. 117). Meshworks are characterised by movements taking place simultaneously, as opposed to practices of merely transiting and transitioning between access points. Consequently 'time and space are not separable within a texture so conceived: space implies time, and vice versa. These networks are not closed, but open on all sides to the strange and the foreign, to the threatening and the propitious, and to friend and foe. As a matter of fact, the abstract distinction between open and closed does not really apply here' (Lefebvre, 1991, p. 118). Sassen uses the term 'assemblages' which she describes as partial and often highly specialised formations centered in particular utilities and purposes (Sassen, 2006a). Meshworks and assemblages differ from network infrastructures. Networks tend to be more similar to point-to-point connections between completed, fixed objects, assembly-chain-like in nature, based on principles of fragmentation, instrumental orientation, and centralised planning. Meshworks, in contrast, are traces of activity in a network that are contingent and that have a weave that is textured by patterns of everyday social activity.

Loosening the grip of space and time

Current infrastructure still requires information to be served from somewhere and delivered to somewhere; at geographic scales a 'bit' of information always has an associated location in a real geographic space. The nodes and networks of mobile and wireless technologies may appear to be invisible and placeless, but the technologies that enable and access them are located in social and physical spaces; where people are, what they're doing and whom they're communicating with. Thus the spaces through which we move become visible in terms of their network accessibility, and consequently, in terms of their implied spatial locality. But as these infrastructures enable new kinds of connections they set up new assemblages. Accordingly, these new infrastructures can 'only liberate activities from their embeddedness in space by producing new territorial configurations, by harnessing the social process in a new geography of places and connecting flows' (Swyngedouw, 1993, p. 306). The consequence of this is that 'networked technologies will also, and in fact, already are, leading to new forms of urbanity' (Sassen, 2011).

One of the key ways that urbanity is being reconfigured by networked infrastructures is linked to the phenomena characterised by Cairncross as the 'death of distance' where communications technologies 'loosen the grip of geography' (1997, p. 5). This is manifested in a range of ways. The first is what Mitchel termed 'tunnel effects' where by 'new spatial patterns emerge (or additional patterns to re-emerge) by erasing incompatibilities, radical reductions in interactive costs can generate profoundly anti-spatial interdependencies between towns, cities and regions' (2002, p. 411). This condition 'generates dramatic slippages and discontinuities in the urban fabric' (Mitchell, 2002, p. 411). Global financial centres are a key example, where a high-rise block in the centre of town operates as a highly connected node in global financial networks, but can be surrounded by social housing areas with high unemployment. Sassen further argues that the proliferation in these globally orientated but locally operating activities means that 'context the city becomes a strategic amalgamation of multiple global circuits that loop through it … and this produces a specific set of interactions in a city's relation to its topography' (Sassen, 2006b, p. 4). The new urban spatiality means that what happens in cities is only partly taking place in what we might conceive of as the space of the city; whether this is literally its buildings and organisations or what people imagine being the city. In this way 'if we consider urban space as productive, as enabling new configurations, then these developments signal multiple possibilities' (Sassen, 2006b, p. 4). As Graham points out 'rather than simply substituting or revolutionising the city, and flows of people and material goods, the evidence suggests that new technologies actually diffuse into the older urban fabric offering potential for doing old things in new ways' (Graham, 1997, p. 173).

Changing temporal infrastructures also loosen the grip of time on space. The flows of materials, goods and people move between time zones and across borders. Sheller and Urry (2006), in their work on mobilities highlight how 'urbanism has always been associated with mobilites and their control, and continues to be so more than ever. The technologies, infrastructure, material fabric and representational machinery of cities support these mobilities, whilst also being shaped and re-shaped by them' (2006, p. 2). An example of how shifting temporal and spatial structures are affecting everday lives is the more fluid relationship between the frameworks that define home and work. In the post industrialised world, the activities that constitute work, home, entertainment and production are less stable as fixed and distinct locations. The possibility for connections to be made either by staying still or by moving between many places 'are transforming the mix of activities within the home, office and automobile in ways that are only beginning to be recognised and understood' (Moss and Townsend, 1999, p. 31). Urry argues that 'networks are viewed as person-to-person connections (whether or not they are sometimes face-to-face). But this ignores the 'material worlds' that organise and orchestrate such networks' which include 'infrastructures of transportation' (Urry, 2003, p. 161) infrastructures of telecommunications (Graham and Marvin, 2001) and infrastructures of mobility (Brown, Green and Harper, 2002). The death of distance argument that 'naively assumed that telecommunications and transportation were substitutes' (Townsend, 2003, p. 63) is refuted by the increasing rise of local and global travel. In fact telecommunications created more demand for travel, as it made it easier to manage global alliances and enterprises. Instead there is a closer linkage between sites of transportation; airports, train stations and buses as part of communication infrastructures. In fact, 'even the optic fibres within and between cities, which carry the bulk of the exploding range of electronic communications, are being laid along rights of way and conduits that tend to closely parallel infrastructural systems for physical movement' (Marvin and Graham, 1994, p. 31). Network infrastructure may be loosening the grip of geography, but the inherent logic and requirements of infrastructures mean that they need to weave closely together with the social and material assemblages of urban fabric. The pattern of footfall on the streets of 22@ is mirrored by a flow of data, water, energy and waste beneath the surface.

1.3   HISTORICAL CONTEXT

Technical infrastructure is constructed through and within social infrastructures. According to Star 'it wrestles with its installed base and inherits strengths and limitations from that base. Optical fibres run along old railroad lines, and failing to account for these constraints may be fatal or distorting to new development processes' (1999, p. 382). Infrastructure is typically seen as the technical 'stuff' that connects nodes within a network, and manages flows through various channels and pipes. But these are embedded and 'sunk into and inside other structures, social arrangements and technologies' (Star, 1999, p. 381). Part of the process of untangling the way that network infrastructures are affecting the urban fabric also necessitates understanding that this process has a long history. The rise of the network society is closely tied with certain developments in the nineties and the rise of the Internet. But network infrastructure has its roots much further back with the introduction of wired communications networks. These developments are closely linked with patterns of urban development, since it can be seen that technologies change the pattern of human use of buildings and the city. In this section we consider the historical context of different kinds of infrastructure and explore how networks entangle with urban infrastructures.

Technical infrastructure: from pipes to fibre

The telephone and telegraph infrastructure were introduced along the lines of existing pipes; with wires taking over. Alexander Graham Bell, often credited with the invention of the telephone, foresaw how a wired network could draw on existing utilities infrastructure. In 1878 he published a prospectus outlining how 'at the present time we have a perfect network of gas pipes and water pipes throughout our large cities. We have main pipes laid under the streets communicating by side pipes with various dwellings, enabling the members to draw their supplies of gas and water from one source. In a similar manner it is conceivable that cables of telephone wires could be laid underground or overhead, communicating by branch wires with private dwellings, country houses, shops etc.' (Fagen, 1975, pp. 22–23). Critically, Bell foresaw the potential of this for the city and highlighted how by 'uniting them through the main cable with a central office where the wires could be connected as desired, establishing direct communication between any two places in the city' (de Sola Pool, 1998, p. 187). Bell called this a 'grand system'. According to de Sola Pool, in his study of the impact of the telephone, the rise in urban density that was enabled through the skyscraper and all the vertical buildings in the city would have been impossible without Bell's grand system of the telephone (1977). De Sola Pool outlines how office life in the early twentieth century relied on messages that were traditionally carried by messengers, who used elevators to move between sender and receiver. As businesses expanded, the use of the telephone to automate the transmission of information made it possible for buildings to be connected vertically.

With the transition from a fixed, wired infrastructure model in the nineties to an untethered hybrid of wired and wireless systems in the early years of the 21st century, digital network infrastructure enabled new patterns of linking places and people. The next shift in infrastructure that would have an impact in a similar way to the telephone was the shift from an analogue to a digital communications system and the rise of the Internet. This grand system became the framework on which telecommunications networks of the Internet were mapped. In a 2001 interview with Paul Baran, credited as the inventor of the digital Internet, he points out that 'the problem was that the telephone system was centralised. You had a hierarchical switching system. I figured there was no limit on the amount of communications that people thought they needed. So I figured I'd give them so much communications they wouldn't know what the hell to do with it. The first realisation was that it had to be digital, because we couldn't go through the limited number of analogue links. We built a network like a fishnet' (Brand, 2001). Key to the fishnet was 'the realisation that by breaking the physical address from the logical address you could move around the network and your address would follow you' (Brand, 2001). These new configurations shifted the association of communications infrastructures with utility and transport networks. With the rise of wireless systems such as mobile phones and satellite networks the idea of infrastructure as wires that still physically connected one place to another became more problematic. In the 1990s the mobile phone overtook the traditional role of the fixed line telephone connection. Alongside this a massive backbone infrastructure of cell phone towers grew year by year. In the USA in the 1990s there were fewer than 10,000 cellar towers, but the number of sites proliferated and by 1996 – when the Telecommunications Act exempted mobile communication providers from local zoning requirements – there were 30,000 and by 2000 there were over 100,000 (Wikle, 2002, p. 46). According to the Antenna Search website, as of early 2013 there are over two million antenna sites in the United States (Wiig, 2013). Within the short space of three decades, the infrastructure that enabled the telephone to be freed from its fixed connection became ubiquitous. Technically, the network of the Internet, as opposed to the network of the telephone is a structure that re-assembles a centralised system into a series of links that represented a more or less stable structure. A distributed net of connections superseded this, where each of the nodes operated more or less as a centre that could be reconfigured in real-time.

Social infrastructure: from telephone operators to call centre agents

The telegraph and the telephone initially worked on a structure that involved local exchanges or telegraph offices which were operated by people who mediated the transfer of information from far to near. But as the system expanded rapidly in scale during the first part of the twentieth century, so this network of local connections was superseded. In fact, by the end of 1920s there were 21 transatlantic cables and 3500 other wires under the world's waters, which created an international telegraphy network. A key consequence was that 'communication has broken loose from the need to be carried somewhere by someone' (Wellman, 2001, p. 19). One of the outcomes of this proliferation of non-local connections is that the space became 'more fungible for communication purposes' (Gottmann, 1977, p. 307) as digital networks reworked the pattern of connections related to space. According to Gottman 'the telephone, which made possible a quasi-instant connection between people located at a distance from one another, seemed destined to modify the relationships built into society by distance and the partitioning of geographical space' (Gottmann, 1977, p. 305). This brought into question the role of a city as a place where people came together to meet and exchange goods and information. If we consider the traditional urban settlement in terms of the concept of the polis: it is a city, a city-state and also citizenship and body of citizens. Polis was not understood as a territorial grouping so much as a religious and political association: while the polis would control territory and colonies beyond the city itself, the polis would not simply consist of a geographical area. The pattern of settlement is determined by human need to communicate with others and to obtain information quickly. Mitchell points out the significance of the changing infrastructure, where 'the Net has a fundamentally different structure, and it operates under different rules … . It will play as crucial a role in twenty-first-century urbanity as the centrally located, spatially bounded, architecturally celebrated agora did in the life of the Greek Polis' (1995, p. 8). The rise of the decentralised network meant that the link between the city as a social centre no longer relied on the spatial organisation of the polis. Yet these simple binaries belie a more complex global/local relationship. Townsend argues that the idea of urban dissolution is flawed and 'advances in telecommunications and information technology actually increased the need for institutions' (Townsend, 2003, p. 63). Instead social centres became multi sited and 'much of what we keep representing and experiencing as something local – a building, an urban place, a household, an activist organisation right there in our neighbourhood – is actually located not only in the concrete places where we can see them, but also on digital networks that span the globe' (Sassen, 2006b, p. 23). The modern day switchboard operator is the call centre agent. In 2002 approximately 3 per cent of the UK and US working population were employed as one. Much bemoaned by many, call centres are seen as telephone connections that lack the sticky relationships and encounters of a face-to-face experience. Call centres are the realisation of decentralised service infrastructures, where the interpersonal contact between people and services is mediated by a remote social encounter.

1.4   MAKING SENSE OF THE CITY

When infrastructure becomes visible

The 'grand system' of integrated infrastructure imagined by Bell is at the heart of our urban infrastructures, and yet, for the most part we are unaware of it. Network infrastructures are 'by definition invisible, part of the background for other kinds of work' (Star, 1999, p. 380). The visible city as a prime determinant of the urban is an artefact of the past. In urban planning there exists a long history of using visual structure as an underlying fundamental of city design. The very idea of the urban plan is grounded on the basis that if a city can be designed to be coherent in layout it will be a successful and inhabitable settlement. Yet networked infrastructures do not have a structural pattern that can be understood in a similar framework to an urban plan. Instead of an organisation that can be considered as a whole, constituted of elements such as buildings, streets and parks, in a networked infrastructure 'the whole city feels like a set of particular points suspended in a vacuum, similar to a bookmark file of Web pages' (Offenhuber and Ratti, 2012, p. 6). The city as a network of connections is not a new concept; work on the value of urban street structure such as that outlined by Hillier and Hanson (1984) prioritises a spatial order that uses street structure as a core unit of analysis. Similarly Lynch's concept of 'image-ability' stated that for a city to be more fully experienced the legibility or intelligible elements of the city need to be understood (Lynch, 1967, p. 4), and found that paths or streets were one of the core elements, alongside landmarks, nodes, districts and edges. According to Lynch, 'a highly imageable (apparent, legible or visible) city would be well formed, distinct, remarkable; it would invite the eye and the ear to greater attention and participation. Such a city would be apprehended over time as a pattern of high continuity with many distinctive parts clearly connected' (1967, p. 10). Lynch's aim was to promote legibility as 'the ease with which [the city's] parts can be recognised and can be organised into a coherent pattern' (1967, pp. 2–3). Indeed the ability of a citizen to comprehend and act upon urban structure through the process of 'seeing' and understanding it is still seen today as one of the core components of urban design. Yet, network infrastructures are not only difficult to conceive visually; they are both immaterial and hidden, they also do not at any point become a coherent whole either spatially or temporally. Critics of Lynch have highlighted how his theory values the visual as the primary method of city structure, but more broadly the underlying concept is that a city can be organised and somehow maintain an organisation that can be understood as a whole with constituent parts. Digital networks, by their nature, are real-time, they come together at nodes and through connections but are not a whole and although at a global scale they can be planned in terms of infrastructure at a local level they are generally invisible and immaterial in a spatial sense. According to Star 'infrastructure becomes visible when it breaks: the server is down, the bridge washes out, there is a power blackout' (1999, p. 382). On 1 August 2014, Facebook went down for a number of hours, and according to media reports police in California got so frustrated with getting calls about the outage that the Los Angeles County Sheriff's Department's public information office was forced to take to Twitter to tell people off for wasting police time (Thomson, 2014). The point at which an infrastructure breaks down or goes down is the moment at which its users become aware of its presence. But this awareness also reveals its lack of transparency and the immutability of infrastructure.

Black boxing

In 2000 I happened to be working on a project with a specialist-engineering contractor whose core area of work was manufacturing and installing mobile phone towers. The contractor had a good deal of knowledge about how to build towers in a range of locations, and worked for all the main UK mobile phone companies. One day we looked through a portfolio of previous projects, and in amongst rather sleek-looking towers were a range of masts camouflaged to look like everyday vertical structures in the environment. Masts were disguised as telegraph poles, complete with false wood graining and there were masts that were created to look exactly like tall pine trees. Mobile phone infrastructure is successfully hidden in a range of ways; flag poles, chimney pots and also church spires; the angel on top of the cathedral in Guildford, UK has a mast hidden under a fibreglass section of his robe (Ward, 2002). It is the same situation with routers that power the Wi-Fi network. It is rare that a router in a public place is visible or marked so as to be recognisable. So the invisibility of much of our ICT infrastructure is not just a case of it disappearing into its environs. The commercial providers who install the infrastructure go to considerable lengths to conceal it, mainly as a result of public concerns about the visual or technological intrusion into the urban or rural landscape. It's not just that infrastructure weaves its way into the background, we actively choose to avoid it, and where possible we expect or even demand that it is concealed.

In the 1930s Lewis Mumford (1934) made a plea for the need to create overarching and historically-informed treatments of the interplay of cities, mobilities and technologies. The idea that lies behind this is that by making an infrastructure visible, it becomes understood as part of a process of integrating it more meaningfully into the structure of everyday life. Sassen's solution to this is to that the major infrastructures in a city – from sewage to electricity and broadband – 'should be encased in transparent walls and floors at certain crossroads, such as bus stops or public squares. If you can actually see it all, you can get engaged. Today, when walls are pregnant with soft- and hardware, why not make this visible?' (Sassen, 2012, p. 14). Yet, the situation is in fact almost the opposite. We increasingly 'black-box' network infrastructure, and hide it from sight. In fact, when infrastructure is by its size and location visible in urban space many attempts are made to deliberately conceal it. Similarly, if you asked most people to point out where the nearest mobile phone mast, router or data center is they would find it difficult to give a clear answer. Yet in urban centres these infrastructures are increasingly densely located. The fact is they are not just invisible (many utilities are hidden but this doesn't generally prevent us from understanding how they work), but are actually hidden and not allowed to be accessed by the public. According to Easterling 'few would look at a concrete highway system or an electrical grid and perceive agency in their static arrangement.

Agency might only be ascribed to the moving cars or the electrical current. Spaces and urban arrangements are usually treated as collections of objects or volumes, not as actors. Yet the organisation itself is active. It is doing something, and changes in the organisation constitute information' (Easterling, 2012). One of the features of meshworks and assemblages is that they infer the presence and participation of different actors, including people that in some way can affect the way that the infrastructure is realised. If infrastructure needs to be transparent (which is different from visible) in order for us to make sense of it in the context of urban fabric, then we need to start finding better ways to reveal the social relations that give it agency.

Air, sky and roofs

Infrastructures may be hidden beneath the surface of the city; in conduits and ducts, but they are also increasingly dominating the air and the sky. The air around us has become a major infrastructural space for networks. Most telecommunications network use air as a form of flow at some point during its transmission. Although we require satellites, antennae's and devices to send and receive data, the local space in which the data is transferred is not confined to a definable route. The infrastructure is literally invisible, and in fact the material within the infrastructure is so widely distributed through packet networks that it does not actually exist in a way that we can identify as material. Recent projects such as Project Loon by Google (Google, 2014) and Facebook's Ascenta drone program (Zuckerberg, 2014) have sought to use large balloons and drones to create a lightweight and flexible infrastructure to broadcast Wi-Fi signals on a large scale. Such projects show that telecoms companies are increasingly viewing airspace as a new infrastructural zone. This is on top of the approximately one thousand satellites that orbit at between 20,000 and 40,000 kilometres above the earth that provide key infrastructure to support television, telephone and navigation systems.

The point at which buildings meet the sky is also taking on new status as a specialised site of infrastructure. The roof is these days often a neglected area of a building, but the infrastructures of our telecommunications and power networks increasingly occupy these spaces. Look up and you will see all sorts of antennae protruding from rooftops; the new church spires of a connected society. Roof spaces work because these infrastructures that connect through waves, generally need to have Line Of Sight (LOS) for them to work optimally. Anyone who has tried to connect a laptop to a wireless router or find a mobile phone signal in a network blackspot will have had to negotiate the tricky positioning of the device so that it can 'see' the router or mast. Despite the fact that many infrastructures are either inherently or intentionally made invisible, the underlying technical connections require direct line of sight, since technologies literally work because they are visible to one another. So we have the rather confusing situation where commercial providers go to considerable effort to conceal infrastructures; and yet they need to be visible to each other to work. From the point of view of the city user; we can look up at the sky or down at the ducts beneath our feet if we want to make sense of the infrastructures, but they are not working on the same frameworks as the visual frameworks of the city.

1.2   Google data centre at Mayes County, Oklahoma (image © 2014 Google).

1.5   CASE STUDY: THE CLOUD

When the internet first developed it was referred to as a place – cyberspace (Gibson, 1995), the global village (Mcluhan, 1964; Wellman, 1999), or chatrooms; it was where you went to go online. As the Internet has grown the ability to comprehend it as a distinguishable infrastructure has lessened. The Internet has become so large and so ubiquitous and so intertwined with all aspects of the way we engage with the world that it is difficult to place 'where' it is. Increasingly the way we interact with networked infrastructure is through a concept known as the Cloud. In the following case study I look at the actual footprint and material infrastructure of the Cloud; which in fact consists of vast and numerous data centers located at various locations around the world (see Figure 1.2). These data centers are both primarily invisible to most Internet users, and yet consume vast amounts of energy and resources.

The first rule of data centers is: don't talk about data centers

The term 'Cloud Computing' first emerged around 1997 (Chellappa, 1997) and according to Armbrust et al. 'cloud computing refers to both the applications delivered as services over the Internet and the hardware and systems software in the datacenters that provide those services' (2009). Cloud computing is expanding exponentially each year (it is predicted that that cloud computing will grow to be worth $121 billion dollars by 2015: a twenty six per cent compound annual growth rate from the $37 billion value in 2010 [MarketsandMarkets, 2014]). The commercial use of the term 'the Cloud' employs, what could be seen as deliberately ambiguous language to describe something that actually comprises of a massive technical infrastructure. More literally it is actually large shed-like buildings housing massive amounts of computer servers linked through a network of fibre optic cables. According to Rich Miller of industry magazine, Data Center Knowledge; 'there is no cloud – it all lives in data centers.

The only thing that changes is the name of the data center operator' (Miller, 2014). Or as Michael Manos, who was then Microsoft's general manager of data-center services explained; 'in reality, the cloud is giant buildings full of computers and diesel generators... . There's not really anything white or fluffy about it' (Vanderbilt, 2009). The reality of the infrastructure of the Internet is that it is housed in data centers that contain our digital information footprint; and linked by a vast array of terrestrial and submarine fibre-optic cabling that transmits information. Different internet services run through geographically dispersed locations; and as well as the main data centres there is 'at least one mirror data center somewhere else – the built-environment equivalent of an external hard drive, backing things up' (Vanderbilt, 2009). This means, that not only is a great deal of the world's computing storage and transactions housed in data centres, in fact there is a duplicate (and sometimes triplicate) copy of all this information; and the corresponding doubling of buildings and equipment.

In the relatively short timescale of the last ten years the vision of John Mccarthy in 1961 where 'computing may someday be organised as a public utility just as the telephone system is a public utility' (Garfinkel, 2011) has been realised. The growth and extent of the rise of mega-data centers has been compared to a new Industrial Revolution (Carr, 2009). Just as emerging industries, once powered by water wheels, were by the 20th century able to 'run their machines with electric current generated in distant power plants' (Carr, 2009, p. 11) so data centers function like a utility, a distant but ever-accessible cloud of services. The actual footprint of the Cloud's geographical and physical location is difficult to pin down. Generally, any company that uses data centers does not reveal the location of where its data is housed. Recently Google became the first major company to ostensibly provide the 'locations' of its data centers, although there is in fact no map, and actual addresses are not given (Google). They are ambiguously referred to as 'Oregon' data center or 'Council Bluffs' or even more generally 'Dublin'. Google maps of data centers don't show a pinpoint location, just a vague idea of the town or city they are located within. Even the 'streetview' option shows, not the outside, but inside views of the particular data center, which is fairly pointless as data centers are remarkably homogenous internally. The only sure way I have found to locate a data center is to trawl Google maps in satellite view and spot the massive buildings by their pristine, blank white roofs (this is no mistake, the white 'cool roof' design is used to reflect heat (Geng, 2015, p. 574) from the array of generators around the building perimeter). To help maintain secrecy, Google typically seeks permits for its data center projects using Limited Liability Corporations (LLCs) that don't mention Google, such as Lapis LLC (North Carolina) or Tetra LLC (Iowa). This is grounded in security concerns. 'It's like "Fight Club" says Rich Miller of Data Center Knowledge "The first rule of data centers is: Don't talk about data centers"' (Vanderbilt, 2009). And indeed there is nothing open or transparent about communications network infrastructure. It's also remarkably inaccessible for people who live and work locally to the massive data center operations. Despite the fact that the construction of data centers costs millions and sometimes billions of dollars and that they consume power that outstrips large towns they need very few people to operate them. Even the largest facilities have created jobs for less than two hundred people, with most of the complex data flows managed remotely from company headquarters or are automated. For example a 2014 Google press release highlights how a four hundred million Euro data centre planned for Eemshaven, the Netherlands will employ a maximum of one hundred and fifty people in a 'range of full-time and contractor roles' (Echikson, 2014). If the example of data centres is anything to go by, a $500 million investment in a data center only equates to one hundred extra jobs locally. This has been referred to as the Solow effect; after his statement in 1987 in an article for *The New York Times* that 'You can see the computer age everywhere but in the productivity statistics' (Solow, 1987). The Cloud may have a physical footprint in the shape of large industrial sheds dotted strategically round the work but they exist in a geographic no-mans land, employ almost no people and visitors are not welcome.

Powering the internet

One of the strongest indicators of the impact of the Cloud and network infrastructure is the amount of power it uses. The US EPA estimates that servers and data centers are responsible for up to one and a half per cent of the total US electricity consumption, or roughly five per cent of US GHG emissions. According to Greenpeace, Equinix, one of the largest data center providers with over 100 data centers spread around the world, collectively consumed 1,830 GWh of electricity in 2012, the equivalent to 162,000 average US homes (Cook, 2014). In 2010 Data centers worldwide consumed an estimated 198.8 billion kWh (Koomey, 2011), which is more energy annually than Sweden (US Department of Energy, 2012). The amount of energy required is growing; from 2000 to 2005, the aggregate electricity use by data centers doubled (Koomey, 2011), and by 2017 it is estimated that the Cloud will use up to two percent of the world's electricity (Cook, 2014). Not surprisingly these figures have had a major influence on the location of Data Centers, which are increasingly sited entirely based on access to renewable energy resources, not based on any practical requirement for proximity to company administrative centres. For example, over the last few years Quincy, a city in Norfolk County, Massachusetts, US, has become an unlikely technology outpost, with five data centers and a sixth under construction. Quincy described as 'farming community in the middle of a desert,' has barely 6,900 residents, two hardware stores, two supermarkets and whose largest building is a grain elevator (Glanz, 2012). The choice of Quincy as the location of Microsoft's two hundred acre data centre, was based entirely in the availability of cheap energy and the proximity to the River Columbia and a series of hydroelectric dams. This has also made it popular for other companies, such as Yahoo, Dell, Microsoft and Yahoo, and according to a *New York Times* article, the data centre power usage overwhelms all nonindustrial electric usage (Glanz, 2012). All residential and small commercial accounts in Quincy consumed an average of 9.5 million watts last year, while Microsoft and Yahoo used 41.8 million watts (Vanderbilt, 2009). The location of Facebook's new 300,000-square foot data centre in Lulea, Sweden is based on its proximity to the Arctic Circle, so that for around eight months of the year, the plant will cool itself. Similarly, one of Google's main European data centers is at Hamina in Finland, which uses seawater from the Baltic Sea for its cooling system (Google Data Center). The new factories of the Internet are not hubs of human activity, but rather factories plumbed directly into energy infrastructures at both a local and a global scale.

Centralisation

Whilst the location of data centers is being driven by their proximity to renewable energy sources, there is a need within the Internet for the carriers to have to physically connect somewhere. In order for the Internet to work in the manner we are all used to, all those individual carriers, ISPs and network operators need to exchange data, or what is known as 'peering'. These centres are known as Internet Exchange Points (IEP), and because the speed of the flow of information is often critical in financial processes they tend to be located near to financial or political centres, to reduce what is called 'latency'. There are a series of other strategically located IEPs in London, Amsterdam, Frankfurt and Washington. The world's largest Internet Exchange Point, which is known as DE-CIX, is based in Frankfurt, Germany, and carries more than 3.2 terabits per second peak traffic.

Getting access to the statistics is fairly easy, finding where all this data is held is harder. In 2013 I arranged to meet the director of DE-CIX and to visit the IEP. In smart offices overlong the river in the harbour region of Frankfurt I met with Arnold Nipper, the director and also one of the founders of DE-CIX. However, visiting the IEP involved a short car drive to a location of a series of anonymous low-rise factory buildings that appeared to have no address or name. They were identifiable by the large clusters of cooling equipment on their roofs. There were no windows. To enter I was required to give up my passport and be ID'd by the head of security. Entering the building involved passing through a Biometric identification airlock revolving door and from the relative messy urban environment into the sterile rackspace of the IEP interior. The racks of loudly humming black boxes, whose activity was literally inscrutable, carry the data that enables the likes of Paypal to exchange payment details with Ebay and for banks to undertake transactions on the stockmarket. The nature of the stock exchange is such that much of current trading is based on algorithms that compete in the race to make deals in increasingly fractional spaces of time. And to do so the latency in wired connections can make a difference to the speed of the connection. So financial competitors choose to take rack space adjacent to one another in what are known as 'meet me rooms' which provide 'proximity hosting'. 'It used to be that things were done in seconds, then milliseconds' (Vanderbilt, 2009), but by moving into co-located data centres, reducing latency in the connection by precious fractions of milliseconds. These are not immaterial concerns; it is estimated that a 100-millisecond delay reduces Amazon's sales by one per cent (Vanderbilt, 2009). But despite the economic value of these financial transactions they do not represent the core volume of traffic on the networked infrastructure. Almost forty per cent of the data that runs through these exchanges is streaming content, primarily video and also images. In fact in 2013, two video streaming sites; YouTube and Netflix accounted for almost a half of internet connections by volume in USA (Solsman, 2013). As we become increasingly reliant on the Internet for real time entertainment this means more data centers and more connections. Our digital connections will result in more and more black boxes housing our computing power.

1.6 SUMMARY

The innocuous building at the centre of the 22@ district in Barcelona; the Tanger power plant actually turns out be more important than it looks. When we unpick the ways that the assemblages of network infrastructures interweave with other networks, then it transpires that energy networks and data networks are now tightly linked. In this chapter we looked at how various ways of assembling the city are understood and how network infrastructure adapts to these contexts. Network infrastructures are made up of components; both software and hardware, pieces of electronic equipment, cables, pipes and the service infrastructure that keep it working. The point at which they are recognised as infrastructure is the way in which they are organised or connected into some form of complex framework operating at a number of scales simultaneously. But at the heart of this way of thinking about infrastructure is the fundamental anomaly that a 'whole' is never realised; the meshworks or assemblages are also in the point of being created and recreated. If we are to be able to make sense of the city then to some extent we must come to 'know' how these structures work and how they are realised spatially. This is about a level of translucency, not in the visual sense, but of a level of understanding that enables us to see the various parts of a system and how they connect. What the case study of data centres shows is that not only are data centers and other network infrastructures 'black-boxed' and made technically and visually inaccessible within urban structures, they are also almost without exception lacking in any form of social context. Under the guise of security concerns, they are one of the most inaccessible building types; located in remote, non-urban settings and protected with incredibly high levels of security. But most tellingly the massive physical, data and energy footprint of these infrastructures requires no people. The assemblages have almost no social connections or capacity, only technical. This raises clear challenges for whether we will choose to find ways to integrate and recognise these network infrastructures into the fabric of the everyday life of the city, or whether we will continue to disassociate and deny the presence of the ever growing black boxes of our reliance on network assemblages.

REFERENCES

Ajuntament de Barcelona (n.n.) *22@Barcelona, the innovation district.* http://www.22barcelona.com/documentacio/22bcn_1T2010_eng.pdf: Ajuntament de Barcelona.

Ajuntament de Barcelona *22@ Infrastructure Plan.* http://81.47.175.201/project-protocol/index.php/22-infrastructure-plan (Accessed: 11 August 2014).

Armbrust, M., Fox, A., Griffith, R., Joseph, A., Katz, R., Kon-winski, A., Lee, G., Patterson, D., Rabkin, A., Stoica, I. and Zaharia, M. (2009) *Above the Clouds: A Berkeley View of Cloud Computing.* Berkeley, CA: University of California, Berkeley.

Boyer, C. (1999) 'Crossing CyberCities: Urban Regions and the Cyberspace Matrix', in Beauregard, R. and Body-Gendrot, S. (eds), *The Urban Moment: Cosmopolitan Essays on the Late-20th-Century City.* London: Sage, pp. 51–78.

Brand, S. (2001) interview with Paul Baran. *Wired.*

Brown, B., Green, N. and Harper, R. (eds) (2002) *Wireless World: Social, Cultural and Interactional Issues in Mobile Communications and Computing.* London: Springer-Verlag.

Cairncross, F. (1997) *The Death of Distance: How the Communications Revolution Will Change Our Lives.* Boston, MA: Harvard Business School Press.

Carr, N. (2009) *The Big Switch: Rewiring the World, from Edison to Google.* New York: W.W. Norton and Company.

Castells, M. (1996) *The Rise of the Network Society: The Information Age: Economy, Society, and Culture.* Vol. 1. Oxford: Blackwell.

Chellappa, R. (1997) 'Intermediaries in Cloud-Computing: A New Computing Paradigm'. *INFORMS Annual Meeting.* Dallas, TX: 26–29 October 1997.

Cook, G. (2014) *Clicking Clean: How Companies are Creating the Green Internet* http://www.greenpeace.org/usa/Global/usa/planet3/PDFs/clickingclean.pdf: Greenpeace USA.

Cosgrove, D. (1996) 'Windows on the City'. *Urban Studies*, 33 (8), pp. 1495–1498.

Crang, M. and Graham, S. (2007) 'Sentient cities: ambient intelligence and the politics of urban space'. *Information, Communication Society*, 10 (6), pp. 789–781.

Easterling, K. (2012) 'An Internet of Things'. *E-Flux.*

Echikson, W. (2014) 'Expanding our data centres in Europe'.

Fagen, M.D. (1975) *A History of Engineering & Science in the Bell System, The Early Years (1875–1925).* Bell Telephone Laboratories, Inc.

Garfinkel, S. (2011) 'The Cloud Imperative. Business Report.' *MIT Technology Review.* http://www.technologyreview.com/news/425623/the-cloud-imperative/ (Accessed: 1 October 2014).

Geng, H. (2015) *Data Center Handbook.* Hoboken, NJ: John Wiley and Son.

Gibson, W. (1995) *Neuromamcer.* London: Harper Voyager.

Glanz, J. (2012) 'The Cloud Factories: Data Barns in a Farm Town, Gobbling Power and Flexing Muscle'. *The New York Times.* http://www.nytimes.com/2012/09/24/technology/data-centers-in-rural-washington-state-gobble-power.html?pagewanted=all, 23 September 2012.

Google (2014) *Loon for all: Balloon-powered internet for everyone.* http://www.google.com/loon/ (Accessed: 1 October 2014).

Google Data Centers *Hamina, Finland*. http://www.google.co.uk/about/datacenters/inside/locations/hamina/ (Accessed: 1 October 2014).

Google Data Centers *Locations*. http://www.google.co.uk/about/datacenters/inside/locations/ (Accessed: 1 January 2014).

Gottmann, J. (1977) 'Megalopolis and Antipolis: The Telephone and the Structure of the City', in de Sola Pool, I. (ed.), *The Social Impact of the Telephone*. Cambridge, MA: The MIT Press, pp. 303–317.

Graham, S. (1997) 'Cities in the real-time age: the paradigm challenge of telecommunications to the conception and planning of urban space'. *Environment and Planning A*, 29 (1), pp. 105–127.

Graham, S. and Marvin, S. (1996) *Telecommunications and the City*. London: Routledge.

Graham, S. and Marvin, S. (2001) *Splintering Urbanism: Networked Infrastructures, Technological Mobilities and the Urban Condition*. London: Routledge.

Herman, R. and Ausubel, J. (1988) *Cities and Their Vital Systems: Infrastructure Past Present and Future*. Washington, DC: National Academy Press.

Hillier, B. and Hanson, J. (1984) *The Social Logic of Space*. Cambridge: Cambridge University Press.

Koomey, J. (2011) *Growth in data center electricity use 2005 to 2010*. Oakland, CA. Available at: http://www.analyticspress.com/datacenters.html.

Latour, B. (1993) *We Have Never Been Modern*. London: Harvester Wheatsheaf.

Latour, B. (2011) 'Networks, Societies, Spheres: Reflections of an Actor-Network Theories'. *International Journal of Communication*, 5, pp. 796–810.

Latour, B. and Hermant, E. (2004) *Paris: invisible city*. http://www.bruno-latour.fr/virtual/EN/index.html (Accessed: 1 October 2014).

Lefebvre, H. (1991) *The Production of Space*. Oxford and Cambridge, MA: Blackwell.

Leon, N. (2008) 'Attract and connect: The Barcelona innovation district and the internationalisation of Barcelona business'. *Innovation: management, policy, and practice*, 10, pp. 235–246.

Lynch, K. (1967) *The Image of the City*. Cambridge, MA: The MIT Press.

Mackenzie, A. (2010) *Wirelessness*. Cambridge, MA: The MIT Press.

MarketsandMarkets (2014) *Cloud Computing Market: Global Forecast (2010–2015)*. http://www.marketsandmarkets.com/Market-Reports/cloud-computing-234.html: PR Web.

Marvin, S. and Graham, S. (1994) 'Privatization of utilities: the implications for cities in the United Kingdom'. *Journal of Urban Technology*, 2 (1), pp. 47–66.

Mcluhan, M. (1964) *Understanding Media: The Extensions of Man*. New York: McGraw-Hill.

Miller, R. (2014) 'Will the Netflix Model Gain Traction? Why Service Providers Should Take Note'. *Data Center Knowledge*. http://www.datacenterknowledge.com/archives/2014/01/07/vc-firm-hires-netflix-infrastructure-guru-will-invest-cloud-enablers/ (Accessed: 1 October 2014).

Mitchell, W.J. (1995) *City of Bits*. Cambridge, MA: The MIT Press.

Mitchell, W.J. (2002) 'E-Bodies, E-Building, E-Cities', in Leach, N. (ed.), *Designing for a Digital World*. London: Wiley Academic, pp. 50–56.

Moss, M. and Townsend, A. (1999) 'How telecommunications systems are transforming urban spaces', in Wheeler, J.O. and Aoyama, Y. (eds), *Fractured Geographies: Cities in the Telecommunications Age*. New York: Routledge.

Mumford, L. (1934) *Technics and Civilization*. San Diego, CA: Harcourt, Brace and Company.

Mumford, L. (1961) *The City in History*. New York: Harcourt Brace.

Offenhuber, D. and Ratti, C. (2012) 'Reading the City – Reconsidering Kevin Lynch's Notion of Legibility in the Digital Age', in Berzina, Z., Junge, B., Westerveld, W. and Zwick, C. (eds), *The Digital Turn – Design in the Era of Interactive Technologies*. Zürich: Park Books.

de Sola Pool, I. (1977) *The Social Impact of the Telephone*. Cambridge, MA: The MIT Press.

de Sola Pool, I. (1998) *Politics in Wired Nations*. New York: Transaction Publishers.

Sassen, S. (2006a) *Territory, Authority, Rights: From Medieval to Global Assemblages*. Princeton, NJ: Princeton University Press.

Sassen, S. (2006b) 'Making Public Interventions in Today's Massive Cities'. *Static*. http://static. londonconsortium.com/issue04/sassen_publicintervensions.php (Accessed: 2014).

Sassen, S. (2012) 'Urbanising technology', in Burdett, R. and Rode, P. *The Electric City Newspaper*. http://ec2012.lsecities.net/newspaper/: LSE Cities. 12–14.

Sassen, S. (2011) 'An interview with Saskia Sassen about "Smart cities"'. http://www. nicolasnova.net/pasta-and-vinegar/2011/07/06/an-interview-with-saskia-sassen-about-smart-cities (Accessed: 13 August 2014).

Sheller, M. and Urry, J. (2006) 'Introduction: Mobile Cities, Urban Mobilities', in Sheller, M. and Urry, J. (eds), *Mobile Technologies of the City*. Oxford: Routledge, pp. 1–17.

Shepard, M. (ed.) (2011) *Sentient City: Ubiquitous Computing, Architecture, and the Future of Urban Space*. Cambridge, MA: The MIT Press.

Solow, R. (1987) 'We'd better watch out'. *The New York Times Book Review*. *The New York Times*.

Solsman, J.E. (2013) 'Netflix, YouTube gobble up half of Internet traffic'. *CNet*. http://www. cnet.com/uk/news/netflix-youtube-gobble-up-half-of-internet-traffic/ (Accessed: 1 October 2014).

Star, S.L. (1999) 'An Ethnography of Infrastructure'. *American Behavioral Scientist*, 43 (3), pp. 377–391.

Star, S.L. and Ruhleder, K. (1996) 'Steps Toward an Ecology of Infrastructure: Design and Access for Large Information Spaces'. *Information Systems Research*, 7 (1), pp. 111–134.

Swyngedouw, E. (1993) 'Communication, Mobility and the Struggle for Power over Space', in Giannopoulos, G. and Gillespie, A. (eds), *Transport and Communications Innovation in Europe*. London: Belhaven Press, pp. 305–325.

Thomson, I. (2014) 'Facebook goes down, people dial 911. Police appeal for calm – yes, seriously'. *The Register*. http://www.theregister.co.uk/2014/08/01/facebook_outage/ (Accessed: 1 August 2014).

Townsend, A. (2003) *Wired/Unwired: The Urban Geography of Digital Networks*. Massachusetts Institute of Technology.

Urry, J. (2003) 'Social Networks, Travel and Talk'. *British Journal of Sociology*, 54 (2), pp. 155–175.

US Department of Energy (2012) *International Energy Statistics.* 2014 (2 October 2014) http://www.eia.gov/cfapps/ipdbproject/iedindex3.cfm?tid=2&pid=2&aid=2&cid=r3,&syid=2008&eyid=2012&unit=BKWH: US Department of Energy.

Vanderbilt, O. (2009) 'Data Center Overload'. *The New York Times*. http://www.nytimes.com/2009/06/14/magazine/14search-t.html?pagewanted=all&_r=0, 8 June 2009.

Ward, M. (2002) 'How mobile phone masts "vanish"'. *BBC News*. Monday, 16 September 2002. http://news.bbc.co.uk/1/hi/in_depth/sci_tech/2000/dot_life/2261039.stm (Accessed: 1 October 2014).

Wellman, B. (ed.) (1999) *Networks in the Global Village: Life in Contemporary Communities.* Boulder, CO: Westview Press.

Wellman, B. (2001) 'Little Boxes, Glocalization, and Networked Individualism', in Tanabe, M., van den Besselaar, P. and Ishida, T. (eds), *Second Kyoto Workshop on Digital Cities II, Computational and Sociological Approaches*. London: Springer-Verlag, pp. 10–25.

Wiig, A. (2013) 'Everyday Landmarks of Networked Urbanism: Cellular Antenna Sites and the Infrastructure of Mobile Communication in Philadelphia'. *Journal of Urban Technology*, 20 (3), pp. 21–37.

Wikle, T. (2002) 'Cellular Tower Proliferation in the United States'. *Geographical Review*, 92 (1), pp. 45–62.

Zuckerberg, M. (2014) Facebook. https://www.facebook.com/zuck/posts/10101322049893211 (Accessed: 13 August 2014).

2

Places

2.1 NETWORKED PLACES

Station

2.1 Screengrab of Foursquare app photos of Paddington Station venue, London (March 2014).

Back in May 2013 I stopped for a coffee in London, whilst in transit through Paddington Station. Paddington symbolises for me the gateway to London; it's the train station that I arrive at when I travel up from my work in Plymouth in South West UK. To mark my arrival I would make a point of making a check-in via the Foursquare app. Foursquare was a location-based social networking app that offered what is termed 'local search' and works by providing information based on people's real-time location from their mobile phone data. The app has changed a bit since 2013, but back then you could register your presence at a Foursquare location through what was known as a 'check in'. By checking in to a venue, you could read tips left by other users about the place and also see in real time who was at the venue and who had visited in the past. So, on this early morning in Spring 2013, coffee in hand, I used Foursquare to check-in to the main station, something that apparently 45,683 people had done in the past, with many of them returning (like me) because this accounted for 152,756 total visits (Foursquare, 2014a). I was made aware that I was not alone on that day, because Foursquare reported that nine other people were also somewhere in Paddington Station at the same time. But what made me rethink how I viewed the station as a place was not just the idea that I was sharing the station with numerous unknown other people. It was the photo stream of uploaded images people had chosen to share of the station. Although there were quite a few pictures which captured the turn of the twentieth century iron roof, the platform and trains which epitomise the traditional idea of a station, there were many images that indicated that the place had other meanings for those who passed through it. Altogether over two thousand nine hundred images, capturing things as diverse as; train tickets, train conductors, meals, selfies, other passengers, crowds, luggage, bags of shopping, taxis, the interiors of trains and departure boards (Figure 2.1). These were interspersed with these were hundreds of pictures of the bronze statue of Paddington Bear, that marks the memorable role of the station in a popular children's story about a teddy bear who is discovered at Paddington and is subsequently named after it (Bond, 1958). For many travellers the station was just a backdrop to a whole set of other activities, people and encounters and events, and for quite a few the station had more importance as the home of Paddington Bear. As well as a multiplicity of places, the station also unravelled as a place that had been recorded at a whole range of different times; tickets marked the dates of journeys, departure boards showed delays that were now long passed and tourists had obviously long since taken the journey home.

My experiences at Paddington train station, mediated through a mobile phone, highlighted a shift in the complex relationship between the city as a place and the city as information. It revealed a more complex intertwining of the two. Mobile social networking and GPS-based apps and interfaces initiate a re-coding of place, moving from map-based and abstract to social and networked; necessitating a re-assessment of how we think about place and space.

Space and place

Paddington Station is both a space; literally a location and a functional designation as a train station, and also a place; as the many socially constructed encounters and activities demonstrate. In order to explore how space and place, and the corresponding space of places and space of flows operate in a networked world I first seek to set out some definitions and ways in which to work with the terms. Space and place are constructs that enable us to make sense of the world. We live and act in a built world; cities, buildings, constructed space. The most focused example of built space is the city. Yet space is also a term that contains a degree of ambiguity that reflects different contexts and experiences. Space can be outer space to an astrologist, it can be an emptiness or a lack of something or it can be a volume to be modelled. In this chapter we work with the term space along the lines of Lefebvre and Massey; as socially constructed. Lefebvre defines space as a social product (1991). A simple interpretation is that a place is a location; a distinct locality or unit of space that can be distinguished from somewhere else. From a phenomenological perspective, a place becomes differentiated from a space as we familiarise ourselves with locations and thus attributes and values (Tuan, 1977). A place is often understood as being constructed through meaning; we talk about a 'sense of place' (Relph, 1984). So, practically, we know what a café looks like, but our favourite café is one that is tied up with distinct and evolving memories, relationships, times and links to other places. Concepts such as space and place are brought into question by networked technologies and infrastructures. Spatialising these communication technologies and thus reconnecting them to spatial settings requires new views on the inter-connectedness of location and behaviour.

Placelessness

Many theorists have used alarming language to highlight the problem of the loss of sense of place as result of network infrastructures and connections. Castells (1998) laments the loss of the 'space of places', which he claims is obliterated by the 'space of flows'. Mitchell claims that 'the classical unities of architectural space and experience have shattered' (1995, p. 44). Augé introduced the term 'non-place'; to refer to places of transience that do not hold enough significance to be regarded as 'places' (2009). Although the apparent 'placelessness' of our urban life is seen as a combination of many factors, urbanisation, globalisation and the rise of networked infrastructures are seen as the main culprits. The consequence of the ubiquity of networked technologies in everyday life is that they uncouple wired and fixed locations that result in a dislocation of person from place.

This raises challenges where, according to Castells; 'in a world marked by abstract flows of information, and characterised by the uprooting of culture and the capture of experience in real virtuality, the marking of spaces, the new monumentality, the new centralities, the attribution of identifiable meaning to the places where we live, work, travel, dream, enjoy, and suffer, are fundamental tasks in reconstructing the unity between function and meaning' (1998, pp. 27–28). Castells tends to assume that function and meaning need to correlate; a construct that may no longer be useful. Forlano provides a more nuanced reading where she argues that; 'it is possible to understand the ways in which social practices and technological actors work together to refute simplistic understandings of space, and rather, mutually constitute meaningful interactions that happen in place' (Forlano, 2013). A station may still be a place to catch a train, but it is also the site of a myriad of different journeys coordinated through online and offline connections, recorded through a range of media and planned and spontaneous meetings that shift through mobile phone calls, emails, face-to-face encounters and even selfies with the stations namesake; Paddington Bear. This is the event of place.

Spatial relations in the network

A wide range of authors such as Hampton et al. (2010), Graham (1997), Ito (2005), Gordon and de Souza e Silva (2011), Aurigi and de Cindio (2008), Coyne (2010), Greenfield (2006), and McCullough (2004) have described how networks and place are not mutually exclusive, and can instead be reconfigured in new and sometimes unexpected ways. Earlier work in sociology and communication theory also provides a context for how communication and place are situated. The approach of Goffman (1963), who highlighted how human behaviours are rule-based 'interaction orders' constructed around situations that use place as a context, was later taken up by Meyrowitz, who tested these ideas in the context of media use and found that 'where one is has less and less to do with what one knows and experiences' (1985, p.viii). Network technologies are intimately related to the spatial world, in that they enable communication at a distance and as such free communication from a fixed location in space. If our conception of space is less about defining actions at fixed places and times, then we see space opening up as what Graham and Marvin refer to as 'relations and processes rather than objects and forms' (1996, p. 414). This prioritises the person-centred, socially and temporally situated view of the city, rather than an objectified structure that we move within. The city therefore becomes a site for mobile technology and associated social practices such as messaging, searching, meeting and tagging.

Aims

In this chapter I move beyond discussion that contrasts real and virtual, hard urban reality and electronic utopias. I start by looking at what constitutes space and place and how we might characterise digital places. In order to explore the ways in which networked place are manifested I highlight a series of changing space-based relations between people, technology and place. This includes an exploration of places in networked infrastructures in terms of them being 'situational'; that is indeterminate and occupied for unspecified periods of time. Through a reflection on the way that technology has changed place-based relations over the last fifty years, it is possible to see how digital interactions rely on socially constructed characteristics of place. What is changing is the increasing mobility and the shift of computer-based communication away from fixed computers to ubiquitous technologies such as smartphones and the associated social networks applications. This results in a merging of physical places and online social spaces, which is manifested most clearly by the increasing importance of inbetween places, and the ability of people to move their attention between online and offline spaces almost seamlessly, but also resulting in a heightened sense of the here and now. This idea is developed in a case study, that focuses on the practices around the use of a location-based social network mentioned earlier in the chapter; Foursquare. Although the app has changed somewhat over the last few years, the case study aims to give an insight into how the experience of place is changed through interaction with a range of digital interfaces and interactions.[1] The study highlights how the experience of places in Foursquare more closely maps inbetween spaces, and suggests that new typologies emerge that differ from traditional functionally defined spatial frameworks.

2.2   DIGITAL PLACES

The space of flows and the space of places

According to Lefebvre (1991) space is produced and supported by social relations. To Proshansky, the physical environment that we construct is more a social phenomena than a physical one (Proshansky, Ittelson and Rivlin, 1970). From an architectural viewpoint architectural and urban space are containers to accommodate, separate, structure and organise, facilitate, heighten and even celebrate spatial behaviour. Space creates settings that organise our lives, activities and relationships (Lawson, 2005). Space is more than just an experiential phenomena – it is also a way of organising the world and framing behaviour. For example, a café is a built space created by an architect and a builder in which we undertake the social activity of drinking coffee, meeting, sitting or working. We might have certain expectations of the space based on our idea of what a café is; where it is located, how it is run, what it might feel like and what sort of activities it supports. Massey gives us a particularly subtle reading of space; which she argues is not a finite entity but is constituted 'through its relations' (Massey, 2005, p. 107). She describes a condition where space 'is always being made and is always therefore, in a sense, unfinished (except that "finishing" is not on the agenda)' (2005, p. 107). Space is comprised of a multiplicity of open-ended, interconnected, trajectories that produce what Massey terms that 'sometimes happenstance, sometimes not – arrangement-in-relation-to-each-other' (Massey, 2005, p.111). Altman and Low (1992) define place as something we establish relations with and delineate place attachment as the 'affective bonds' observed to have been developed between people and places. The focused study of this psychological connection that people develop through experience and express it in relation to a particular location started in the seventies when human geographers like Tuan (1977) and Relph (1984). A later study by Manzo and Perkins found that the length of residency within a location was considered to be one essential component of the place attachment phenomenon (2006). The

---

[1]   In July 2014 Foursquare removed the check-in and location sharing functionality of the app. This essentially shifted the app from a community of users to a location recommendation platform (for further background refer to press articles e.g. Popper and Hamburger, 2014 and Tate, 2014). The current version, Foursquare 8.0 (2015) focuses on offering personalised local searches, giving users different results based on their preferences and activity. However the use of Foursquare documented in this chapter captures a time when the users ability to 'create' places and the nature of the associated sharing practices revealed a whole range of ways on which people made sense of place and its connection with people through mixed online and offline interactions. Although the development of the app has closed these routes to user involvement, studies of the app between 2010 and late 2014 are still valid in terms of a discussion around the changing nature of space and place.

distinction between networked spaces and physical places is often characterised as a dialectic; as highlighted by Castells who states that 'there is a growing tension and articulation between the space of places and the space of flows … cities are structured and de-structure simultaneously by the competing logic of the space of flows and the space of places' (2004, p. 85). The argument is that places are dissolved by flows, the real is negated by the virtual and face-to-face meetings are wiped out by online interactions. On the one hand such network technologies, which whilst crucial in supporting the mobility and flux, are also fixed networks that must be embedded in space and place. They are also now embedded in the practices of everyday life. Networks of flows do not just operate at a global scale, but create local structures and frameworks with local and very much situated effects. According to Gordon and de Souza e Silva 'when spaces are both physical and digital and when interactions between people are mediated, this does not spell the end of good urban spaces; but it does spell a change' (2011, p. 86). The changes that occur can in part be seen as a re-ordering of spatial paradigms or syntaxes (Aurigi, 2012). This includes a reconfiguration of Euclidean spatial frameworks that is fundamentally different from the PC internet (Willis, 2008).

Situations

The Situationists, a semi-political movement that reacted to the commodification of space in Paris in the 1960s, set out a theoretical position together with a series of approaches and actions aimed at re-appropriating spaces in the city. De Certeau (1984) describes a Situationist position on place and space by making a distinction 'between space (espace) and place (lieu) that delimits a field. A place (lieu) is the order (of whatever kind) in accord with which elements are distributed in relationships of coexistence … The law of the "proper" rules in the place: the elements taken into consideration are beside one another, each situated in its own "proper" and distinct location, a location it defines. A place is thus an instantaneous configuration of positions. It implies an indication of stability. Space is a practiced place' (1984, p. 117). The Situationists advocated walking as a spatial tactic, since it 'affirms, suspects, tries out, transgresses, respects etc. the trajectories it speaks' (de Certeau, 1984, p. 99). Massey also uses the idea of trajectories to open up alternative readings of space, where 'space is the sphere of the possibility of the existence of multiplicity' that is space 'as the sphere in which distinct trajectories coexist; as the sphere therefore of coexisting heterogeneity' (2005, p. 9). For Massey, space is always under construction; 'it is always in the process of being made. It is never finished, never closed' (2005, p. 9). Space as a concept must be understood as a multiple, layered setting in which an individual perceives, acts and interacts. In everyday life it is difficult to distinguish between the particular aspects of a framework and the multiplicitous behaviour we display within it; they overlap and interweave in often indistinguishable ways. Thus a place such as railway station may not just act as a place from which to catch a train, but as a space for waiting, for reading, for loitering, for watching, for meeting and more. Critically, these multiple activities are not mutually exclusive; and so, for example, the individual act of waiting does not detract from the ability of the station to host the arrival and departure of trains. In this sense space is incredibly flexible in allowing multiple activities to occur simultaneously without affecting its integrity.

Hight terms this multiplicity of place when experienced with media as 'narrative archaeology' and outlines how media can be used to reveal the narratives embedded in the place; 'our cities, towns and the landscape as a whole can now be navigated through layers of information and narrative, of what is occurring and has occurred as well as mapping of the physical place. Narrative, history and scientific data are a fused landscape, not a digital augmentation, a multi-layered, deep and malleable resonance of place' (2013, p. 251).

Place and urban social life

The urban theorists Jane Jacobs and William Whyte were part of a shift in thinking in the 1980s where the social life of urban places was given value as part of their role in the social capital of the wider community or neighbourhood. Jacobs described the way that place is animated by human activity which she called place 'ballet', where 'under the seeming disorder of the old city, wherever the old city is working successfully, is a marvellous order for maintaining the safety of the streets and the freedom of the city. It is a complex order. Its essence is intricacy of sidewalk use, bringing with it a constant succession of eyes. This order is all composed of movement and change … an intricate ballet in which the individual dancers and ensembles all have distinctive parts which miraculously reinforce each other and compose an orderly whole. The ballet of the good city sidewalk never repeats itself from place to place, and in any once place is always replete with new improvisations' (Jacobs, 2002, p. 50). Whyte highlighted the importance of small urban spaces in the social life of the city, and found that different places created different situations for human behaviour. He valued 'a place where people come together, face-to-face. The [city] center is the place for news and gossip, for the creation of ideas, for marketing them and swiping them, for hatching deals, for starting parades … But this human congress is the genius of the place, its reason for being, its great marginal edge' (Whyte, 1988, p. 341). Oldenburg also points out that neighbourhood identity and a sense of belonging is in 'third places'; those beyond the first and second places of home and work respectively (Oldenburg, 1999). Oldenburg outlines eight characteristics of third places: 'neutral ground, leveller, conversation, accessibility and accommodation, regulars, low profile, playful mood, and home away from home' (Oldenburg, 1999). These are places such as cafés, bars and parks and the transitory meaning people associate with these places (Oldenburg, 1999). This is reflected in the work of Gehl (1987) who also aimed to better understand how people used spaces and places in the context of the social life of the city. According to Gehl 'life between buildings offers an opportunity to be with others in a relaxed and undemanding way. One can take occasional walks, perhaps make a detour along a main street on the way home or pause at an inviting bench near a front door to be among people for a short while' (Gehl, 1987, p. 17). The social life of a community is played out in streets, cafés, parks, stations and other transitory or 'inbetween' spaces where people have access to gathering and encountering others.

There have been a number of authors who have commented on the changing social life of 'inbetween' spaces due to the use of mobile devices (Gergen, 2002; Ling and Yttri, 2002). It is argued that people who are using mobile phones to chat or access the internet are being distracted from the immediate here and now, 'so that they participate less in the social interaction we encounter in our geographical place or community' (Harrison and Stephen, 1999, p. 221). If you've recently spent time or passed through an inbetween space, such as in a station or a café or during a train journey or bus ride, you will have probably noticed the numbers of people occupied by their phones or tablets. The rise in ownership of smartphones has heightened this use, and according to a 2012 report over eighty per cent of owners worldwide use smartphones 'on the go' (Google, 2012). Ito et al. describes this as 'a colonisation of in-between space' (Ito and Okabe, 2005, pp. 263–264). Spaces previously characterised as only useful for passing through: non-places or transit spaces have become sites for a range of digital social practices and encounters. The context of place is still a constructive quality in the use of phones and other mediated interactions in inbetween spaces, but it is no longer the defining factor. The place is still part of the situation, but we don't necessarily need to be at a specific place. For example, Twitter or Foursquare create a 'territory' or media zone in which users construct their own way of being present with others' (Buschauer and Willis, 2013). Places are experienced not as discrete and definable locations, but are understood as a quality of the situation in which we are communicating and acting.

2.3   HISTORICAL CONTEXT

Technological places: from fixed to mobile

Communication across distance was primarily bound up with the speed of different modes of transportation until the turn of the twentieth century. For example, in 1900s transatlantic letters were sent on ships that sailed intermittently, so a letter could take a couple of weeks to arrive. By 1900 the telegraph emerged as a mode of communication enabled by a cable laid under the Atlantic. However it was the telephone that first made significant shifts in the relationship between communication and place. Bell's first successful words on the telephone reveal that he saw it as a technology to pass messages to bring people together in the same place, rather than uniting them across distance. In 1876, his opening instruction was to ask his assistant at the other end of the line to come to where he was: 'I then shouted into M [the mouthpiece] the following sentence: "Mr. Watson--come here--I want to see you." To my delight he came and declared that he had heard and understood what I said' (Bell, 1876). The telephone was still very much a fixed technology, but it enabled significant changes in the mobility of people between places and also broader impacts such as 'increased the spatial distribution of labor and society; the home could be more distant from the place of work' (Gottmann, 1977, p. 312). As a consequence the telephone was seen as the main factor which allowed geographical separation between office work and stages of business, such as production, warehousing and shipping of goods (Daniels, 1975). As we have seen in Chapter 1, the telephone still connected through a fixed wired infrastructure but it meant that places and buildings no longer needed to be in proximity of each other in order to communicate with one another since 'electronic messages seep through walls and leap across great distances' (Meyrowitz, 1985, p. 117). Around the middle of the last century the television also emerged as a key technology that changed the way that information was accessed in relation to place; the television. According to Scannell the television contributed to a 'doubling of place' where public events 'occur simultaneously in two different places, the place of the event and that in which it is watched and heard' (1996, p. 76). Meyrowitz (1985) highlights how this shifted the relationship between built space as a connector and also providing physical separation since 'the walls of the family home ... no longer wholly isolate the home from the outside ... Children may still be sheltered at home, but television now takes them across the globe before parents give them permission to cross the street' (1985, p. 67). Initially these technologies were expensive and took a privileged place and role in the home and office. But as costs dropped, the television and the telephone soon became ordinary technologies. Media started to become less fixed and more portable in relation to place, with the introduction of devices such as the Walkman (a Sony brand trade name originally used for portable audio cassette players) and then in the eighties the mobile phone. The other form of everyday technology; the personal computer started to become accessible to the general public in the eighties, but it was introduced as a fixed technology. In the nineties Mark Weiser, a researcher at Xerox Parc (a research and development company in Palo Alto, California specialising in information technology and hardware systems) set out a future vision in which computing technologies would be 'invisible', a phenomena that he called the third great paradigm after the mainframe and personal computer. This led to thinking about how computers could become more embedded within the world around such that 'they weave themselves into the fabric of everyday life until they are indistinguishable from it' (Weiser, 1991, p. 94). This made the final step in freeing up communication from place. It also opened up more careful consideration of the role of place. In their essay 'Replace-ing Space' Harrison and Dourish argue that designers needed to move from thinking about spaces, to working with places since 'Space is the opportunity; place is the understood reality' (1996, p. 67). They introduce an early idea of 'hybrid space' where 'two people can be what they think of as the same place (like an electronically shared office), but will not be in the same physical space, nor even will they be the same hybrid space. ... Each of us is in a separate space; linked, but not shared' (Harrison and Dourish, 1996, p. 74). Whether online, offline or hybrid the experience of place was becoming shared; shared between people, locations (often mobile), time and types of mediated interaction.

Social places: from MUDS to inbetween spaces

In terms of its impact on place and society, in the early twentieth century, a key feature of the telephone was to connect people who lived in isolated communities; 'the telephone has immensely improved the faculty of isolated people to communicate with others outside their households. To people settled in a scattered pattern, in relatively isolated buildings, the telephone gave a heightened sense of security against hazards and provided an escape from loneliness' (Gottmann, 1977, p. 307). Over the course of the next fifty years the telephone became a ubiquitous technology, present in almost every home or workplace. With the introduction of the Internet, the last twenty years of the twenty first century saw a proliferation in the ways in which people could communicate. As the computer became ubiquitous the concept of the virtual community was introduced by Rheingold (2000) who explored the social implications of the use of online 'places' such as Usenets, MUDs (Multi-User Dungeon) and Internet Relay Chat (IRC), chat rooms and electronic mailing lists. In these communities people were brought together by a common interest, rather than a shared physical location. Initially these online communities were referred to as places, as in Rheingold's description of the WELL, one of the first online communities: 'it's like having the corner bar, complete with old buddies and delightful newcomers and new tools waiting to take home and fresh graffiti and letters, except instead of putting on my coat, shutting down the computer, and walking down to the corner, I just invoke my telecom program and there they are. It's a place' (Rheingold, 2000, p. 9). Rheingold went on to describe the WELL's 'place-like' aspects and compare 'cyberspace' to the equivalent of Oldenberg's 'third place'. He questioned whether 'perhaps cyberspace is one of the informal public places where people can rebuild the aspects of community that were lost when the malt shop became a mall' (2000, p. 10). Wellman, a sociologist who has looked at the effects of communication networks on community, also identifies the fact that there was a consequent shift around the end of the last century from place-to-place to person-to-person community. He highlights this as problematic because he sees that 'compared to door-to-door community, place-to-place community operates in a contextual vacuum' (Wellman, 2001b, p. 235). Wellman highlights how 'place – in the form of households and work units remains important – even if neighbourhood or village does not. ... Households and work units are important bases of interaction. They also provide places from which their automobiles, (wired) phones and Internet connections operate' (2001a, p. 34). In contrast to Rheingold, Wellman finds less examples of the third space phenomena saying that; 'people and places are connected. Yet there is little social or physical intersection with the intervening spaces between households. It is place-to-place connectivity, and not door-to-door. People often get on an expressway near their home and get off near their friend or colleague's home with little sense of what is in-between. Airplane travel and email are even more context-less' (Wellman, 2001a). With the emergence of location-based social networking, where the computer connection is longer linked to a fixed location it is these inbetween spaces that emerge as sites of social interaction. In his study of an online social network iNeighbours, Hampton found that it increased social cohesion in the neighbourhoods where it was used (2007), and that this spilled over into people being more neighbourly locally and thus more connected to the place in which they lived.

2.4   DIGITAL PLACES

If we shift our thinking from a view of place as a distinct and bounded location that has meaning through the relationship we construct with it over time, to the idea of place as socially constructed and part of contingent situations and trajectories this does not mean that places do not still provide meaning in our lives. In this section I investigate how places, as part of contingent situations, are created when social practices and technological actors mutually constitute meaningful interactions. But to do this I try to unpick some of the ways that place is produced, and understand how these characteristics are remapped onto new practices.

Inhabiting the inbetween

Massey (2005) gives value to the 'places in between'. In these space she argues that you are 'travelling not across space-as-a-surface … you are travelling across trajectories' (2005, p. 231). Places becomes part of trajectories that are co-ordinated before, during and after through social networks, such as Twitter, Facebook and WhatsApp. According to Ito and Okabe in their study of Japanese teenagers 'people saw value in residing for a period of time in a desirable location. Just as people seek out beautiful campsites to set out there gear and reside for short periods of time, urbanites find attractive public places to temporarily set up camp with the help of their information technologies' (2005). They term this the 'colonisation of inbetween space' (Ito and Okabe, 2005). Café's, seen by Oldenburg (1999) as one of the key 'third places' are good examples of inbetween spaces where people set up camp, creating a fluid mix of workspace, meeting space and food space. Paradoxically this means that inbetween spaces then start to become less fluid and more stable. Tuters highlights that these practices mean that cafés actually become less transitory since 'todays ubiquitous Starbucks cappuccino bars offer the digital, mobile class a refuge from the pace of city, a space of introspection rather than random encounter. [these places] … form an archipelago of pseudo public spaces throughout the world's cities' (2004). In fact Starbucks welcomes this changing idea of functionality, and announced in a press release in 2009 that 'we do not have any time limits for being in our stores, and continue to focus on making the Third Place experience for every Starbucks customer' (Needleman, 2009).

Certain characteristics make café's attractive as places for 'camping'; the presence of other people and the atmosphere of a comfortable 'home-away-from-home'. Ito et al. in an ethnographic study of the everyday lives of young professionals in London, Tokyo and LA found that 'the attraction of working in a specific "camping site" can include the personal relationships fostered there, food and drink, infrastructures (tables, electricity, Wi-Fi), and most importantly, diffuse social ambience' (Ito, Okabe and Andersen, 2010, p. 78). This list maps closely to Oldenberg's eight characteristics of third places; 'neutral ground, leveller, conversation, accessibility and accommodation, regulars, low profile, playful mood, and home away from home' (1999). The one outstanding feature in Ito et al.'s description is the need for infrastructure. Access to places requires a hybrid of physical 'affordances' (such as a place to sit) and technical properties; inhabiting the place inbetween also means access to power and Wi-Fi.

'Selfie' architecture and inattention

According to Goffman, social interaction 'can be identified as that which transpires in social situations, that is environment in which two or more individuals are physically in one another's presence' (1983, p. 2) and he maintains that the physical environment provides a structure for these encounters. This approach uses presence to measure the meaning of an interaction; it requires a commitment of time and attention. The management of our attention gives rise to the problem of 'absent presence' (Gergen, 2002), where the use of a phone, or viewing a screen means that we are no longer paying attention to our immediate situation; we are in two places at once: 'The other place that she is "on the telephone". And she may well understand that to be a private place. […] (She) is not in the same 'there' as the rest of us are; there are two "theres" there' (Schegloff, 2002, pp. 286–287). This was even proven in a study at Washington University, where they used an interesting technique to test out how much of the world around them people who were using devices were actually aware of. The researchers paid a unicycling clown to move around people in an open public space. Afterwards they asked people whether they had seen the clown. Over seventy five per cent of people who were on the phone said they hadn't seen him, whereas only fifty per cent of the people without phones or who were just listening to their iPod said they had not seen the clown (Hyman et al., 2010). People were physically in the place, but their attention was elsewhere.

But the people on phones are not just somewhere else; they are communicating with someone or interacting with a place remotely. Habuchi (2005) terms this 'telecocooning' where the communication of one person to the next without having physical interaction with that person. Goffman speculated that 'presumably the telephone and mails provide a reduced version of the real thing' (1983, p. 2), and these practices are some examples of how interaction orders are shifting, but they are not necessarily being reduced. The visual sense of presence (being there face-to-face) is increasingly replaced with a presence through communication. Ito et al. found that people used 'mobile phones to bring in the presence of other friends who were not able to make it to the physical gathering, or to access information that is relevant to that particular time and place' (Ito and Okabe, 2005, p. 17). Thus a construction of a shared sense of presence is not achieved through 'being there' or through proximity (Hall, 1966), but rather through the way in which we interact with the space. In fact many of the current social media apps encourage a heightened sense of being present – a form of 'selfie' architecture. Twitter (used by twenty per cent of the US's adult population [Pew Internet Project, 2014]) invites users to publish based on the prompt 'what are you doing now?' and SnapChat works on people sending time-based photos of 'here and now'. They represent a new situationism that brings people's attention into the immediate environment they are in.

Urban legibility and memory

We toggle our involvement and attention from the physical space to media space like an on/off switch. As we toggle inbetween screen and environment, key environmental stimuli are being missed. This raises issues for urban legibility, since this suggests that for an individual to achieve a truly vivid image of the city a series of key elements must be observed and easily understood (Lynch, 1967). But the use of phones and satnav type devices to access information about our surroundings or getting directions means that we often don't pay attention to the spatial world. This is not primarily to do with inattention, but more rooted in the fact that to find our way in a place we need to learn about it. These interfaces act as what is known as 'cognitive offloading' (Hutchins, 1996); they do the hard work of figuring out where we are and how to get somewhere for us, so our engagement with the environment is lessened. A recent article in a British newspaper on 28 August 2014 highlights this point with the following headline: 'Judge blames Sat Nav for cyclists death.' According to the article 'District Judge Roger Elsey told motorist Steven James Conlan that he had paid too much attention to his Sat Nav and not enough to the road ahead when he hit cyclist Grahame McGregor on a crossroads on Easter Monday of that year. Sentencing he said: 'I don't believe the accident would have occurred if the Sat Nav had been switched off.' The driver had been relying on the in-car technology to guide him and his family during a family day out when the Sat Nav failed to register the road junction. The driver drove straight out into the main road, hitting a cyclist who died in hospital five days later' (Tallentyre, 2013). Now we might just say that this was the fault of that particular driver, but there are numerous newspaper and police reports that show that this is a widespread phenomenon (Collins, 2010; GPS Bites, 2012). These newspaper reports and police files document people driving off cliffs, over railway tracks, turning onto the motorway in the wrong direction and driving down one-way streets. When we switch our view to that of the screen, we stop attending to the world. According to Lynch a legible city would allow 'for vivid identification which would in turn extend and deepen ones grasp of the surroundings' (Lynch, 1960: 10,11). But in a portent of what was to come he countered that 'devices are extremely useful for providing condensed data on interconnections they are also precarious, if the device itself must constantly be referred … the anxiety and effort that attend such means 'creates' an experience of interconnection … the full depth of a vivid image is lacking' (Lynch, 1967, p. 11). Consequently, the quality of the visual environment that we tend to think of as important to our construction of the meaning of place;

what it looks like and how this fits within a broader urban setting, becomes less important. But just as in the Foursquare example of Paddington Station; events, time-based changes, remarkable features and other people are memorable, and these can contribute to a 'collective environmental image' of place, a persistently retrievable history of the things that are done and witnessed across any place that can be specified with latitude and longitude coordinates. More generally the features of places also include the way that is experienced, captured and shared with others; they exist in the world of the screen and the world of the eye.

2.5   CASE STUDY: LOCATION-BASED SOCIAL NETWORKS

In this section I test out changing characteristics and structures of places through real examples. This uses a case study of the early use of a location-based social network, Foursquare, to illustrate the changing nature of place. Foursquare places are typically sites of transition and temporal occupation; they are places where people converge and then disperse; brought into being for the time in which networked links connect. The sociality of networked place, how it contributes to community networks is also explored, and consequently how new typologies emerge.

Events and venues

Location-based social networks (LBSN) differ from online social networks (Facebook, Twitter, Instagram etc.) since they work primarily on the actual location-information, so that what you access depends on where you are. According to Zheng 'LBSN does not only mean adding a location to an existing social network so that people in the social structure can share location-embedded information, but also consists of the new social structure made up of individuals connected by the interdependency derived from their locations in the physical world as well as their location-tagged media content, such as photos, video, and texts' (2011, p. 244). In this case study I focus on the use of Foursquare, a commercial LBSN app between 2013 and 2014. In 2009, during the early stages of the app, the Foursquare developers got around the problems of not having an existing location database by utilising their users to solve the problem and enabled user-generated venues. By March 2014, the database had more than sixty million user-submitted venues (Foursquare, 2014b). Because the original version of the app allowed users to create the places, the naming practices that emerged reflect the changing perceptions of how places are understood. Train journeys, events and activities could be observed as the most popular places to check-in, all of which extend well beyond traditional ideas of a discrete location or place. For example, in London the top ten venues are all airports, with Heathrow being the most visited, having a total of 674,838 check-ins and an average of two check-ins per user; the most check-ins at any venue were 1,639,119 at Hartfield Jackson Atlanta Airport, which represented an average of four check-ins per user (Muzychenko and Kats, 2015). Its popularity on Foursquare was probably due to the fact that Hartfield Jackson is statistically the airport with the highest annual passenger numbers worldwide.

In 2008 and 2009 I mapped the names of venues on Foursquare in a German city, where the app was still in its infancy and found 1480 check-ins and 484 users (2 August 2008) (Willis, 2012). The highest proportional number of 'check-ins' was at train and bus stations, airports and on trains, although the University was also high on the list as the city has a large student population. Almost all of the more popular cafés and restaurants were mapped in the app, many of them being take-aways or places where people did not actually sit to eat. As I travelled I used Foursquare to look at naming practices elsewhere. I found that many users registered transit spaces as venues, and it is common practice among Foursquare users to list a train journey (e.g. 'ICE 278 to Berlin') or a flight number (e.g. 'LH 1029 HAJ_MUC') or even roads (M25 Junction 4). The transit space, a previously regarded as a 'dead space' becomes re-valued as a place, where users occupy it just as if they would inhabit a more functionally defined space. These everyday, visually uninteresting places become the foci or datum points through which we orientate our lives.

Mayors and the civic layer

As part of the original gaming format, another feature of Foursquare was that it allowed users to become mayors of places, through regular and repeated check-ins at a particular venue. Schwartz documented how becoming mayor of a venue, such as a regular restaurant or coffee shop gave people a sense of ownership (Schwartz, 2014, p. 92), and gave value to a feeling of connectedness to places that were frequented as part of their everyday journeys or activities. This connection was only manifested in the app and had little relation to the visual or spatial qualities of the actual place, and more to their connection with the other people who worked there or visited. Interestingly the idea of everyday mayors is nothing new, Whyte introduces mayors and checking in from his study of street life in New York (Whyte, 1980), where he found that 'most well-used places have a 'mayor' of sorts. He may be a building guard, a newsstand operator or a food vendor. Throughout the day you will notice people checking in with him – a cop, perhaps, a bus dispatcher, various street professionals, and office workers and shoppers who stop by briefly for a hello or a bit of banter … There may be an older couple looking somewhat confused. He will anticipate their questions and go up to them. Are they by chance looking for a reasonable place to eat? Well, yes that's what they were going to ask him' (1980, p. 313). According to a news article in 2012 Michael Bloomberg of New York also gained a mayor-ship in Foursquare (Undergleider, 2012). But this was an exception and generally, Foursquare mayors are not civic mayors. For example, the Foursquare mayor of London City Hall in 2010 was a digital media specialist called Christine Chau (Rowan, 2011) and the mayor of the US White House in 2011 was a journalist called Aya Maher (Rogin, 2011). So it does not map directly onto existing civic structures. This is also the case for social interaction. In a 2012 study, Humphreys found that Foursquare facilitated parochial places; described by Lofland as 'characterised by a sense of commonality among acquaintances and neighbours who are involved in interpersonal networks' (1998, p. 10). By promoting an awareness of the presence of others, the use of the app 'contributed to the parochialisation of public realm by facilitating person-to-person and person-to-place connections' (Humphreys and Liao, 2013). Humphreys and Liao also describe the how mayor-ships were claimed as a personal territory and were the result of a 'temporal and perhaps financial investment in a particular activity or place' (Humphreys and Liao, 2013). The mayor-ships established a person-to-place commitment, based on time spent there, duration of time at a place and degree of intimacy felt to the place (Humphreys and Liao, 2013). But on other levels, Foursquare mayors were not the equivalent of Whyte's mayors; they were not people to go to for information, they did not approach lost strangers and offer help and in fact a study showed that Foursquare users didn't tend to meet or contact strangers through the platform (Frith, 2012). There are many reasons for this, but it may be that such platforms lack the 'civic layer' (Horan, 2000, p. 62), a fact reflected in the popularity of venues; libraries, town halls, museums and schools tend to be much less popular then stations, airports, shops and sports centres.

New typologies

Foursquare places are tagged with a GPS co-ordinate, so that they have a defined geographical position. But they do not have to submit to standard norms; there can be multiple locations nested in one. For instance, there can be a platform four at Paddington Station, a coffee shop at Paddington Station and also the Paddington Bear statue, all listed as discrete venues. But in early versions of Foursquare, time-based events could also be listed as places. According to a company blog post in 2011: 'It's one of the most common check-ins on Foursquare: you head off to a movie theater, check-in, and type in "Harry Potter" to tell people what you're seeing, or check-in to a stadium and shout "Patriots game" or "Lady Gaga concert." In moments like this, a place is often more than just a place; so today, we're starting to pull major events into our database' (Foursquare, 2011). Foursquare privileges the location of a user; the app focuses on showing what is geographically 'near' to the user, so that you need to be at a location to see other places around you. According to a 2011 study by Cramer et al., participants reported how they felt it was important that they were not 'just passing by, we actually have to be there' (Cramer, Rost and Holmquist, 2011, p. 63). To mark their ten millionth user in 2011 Foursquare released figures about practices of their app users. They report that 'over one thousand births and over six thousand weddings have been published via hospital and church or registry office locations' (Foursquare, 2011). The value and equivalence given to events as places in Foursquare prioritised shared encounters and experiences, not addresses or locations, and this should inform thinking about how we characterise typologies of places that are experienced through on and offline interactions. Most importantly they open up the idea that a space can be an event, and that the inbetween is often where we experience the most meaningful merging of the digital and physical.

2.6   SUMMARY

Digital places are far from placeless, but they do have different characteristics and meanings in our everyday lives to the types of places described by Relph (1984) and Tuan (1977). The flows of information and exchange of networked interactions alter the character and function of space over time. These information flows, social exchanges and encounters become part of the material of the city, just as the height of buildings, the location of parks and the thoroughfares of key streets. But since information flows are ephemeral there is a need to make sense of them just as we make sense of the city through our eyes. The way in which they affect our behaviour makes them distinctive; not necessarily visually, but as meaningful in our everyday social lives. These networked places gain materiality not through the senses but through our attention, which has an impact on the experience of place and also on the nature of urban legibility. We become present and aware of these places as we choose to focus our attention on a mobile screen or a phone call or a text message; a phenomena I characterise as 'selfie' architecture. For this moment we participate actively in a place that shapes how we decide what to do and makes connections with other places, whether near or far. In the Foursquare case study it was shown that remarkable or distinctive places are not realised at unique locations but as events, played out over time and encompassing multiple locations. This privileges inbetween spaces, everyday and often overlooked spaces of transit and social encounters. The experience of digital places evades some of the more formal ways in which we have come to make sense of the physical world around us, but in doing so they open up opportunities for a more situated and socially constructed sense of place.

REFERENCES

Altman, I., and Low, S. (1992). *Place attachment*. New York: Plenum Press.

Augé, M. (2009) *Non-places: Introduction to an Anthropology of Supermodernity*. London: Verso Books.

Aurigi, A. (2012) 'Reflections towards an agenda for urban-designing the digital city'. *Urban Design International*, 18 (2), pp. 131–144.

Aurigi, A. and Cindio, F.D. (2008) *Augmented Urban Spaces: Articulating the Physical and Electronic City*. Aldershot: Ashgate.

Bell, A.G. (1876) 'Lab notebook, March 10, 1876'. [Notebook] http://www.loc.gov/exhibits/treasures/trr002.html: The Alexander Graham Bell Family Papers.

Bond, M. (1958) *Paddington*. London: William Collins & Sons.

Buschauer, R. and Willis, K.S. (eds) (2013) *Locative Media: Multidisciplinary Perspectives on Media and Locality*. Bielefeld: Transcript Verlag.

Castells, M. (1998) 'The Education of City Planners in the Information Age'. *Berkeley Planning Journal*, 12 (1), pp. 25–31.

Castells, M. (2004) 'Space of Flows, Space of Places: Materials for a Theory of Urbanism in the Information Age', in Graham, S. (ed.), *The Cybercities Reader*. London: Routledge, pp. 82–93.

de Certeau, M. (1984) *The Practice of Everyday Life*. Berkeley, CA: University of California Press.

Collins, N. (2010) 'Sat nav mistakes: when technology fails'. *The Telegraph*. http://www.telegraph.co.uk/motoring/news/7931558/Sat-nav-mistakes-when-technology-fails.html, 07 Aug 2010.

Coyne, R. (2010) *The Tuning of Place: Sociable Spaces and Pervasive Digital Media*. Cambridge, MA: The MIT Press.

Cramer, H., Rost, M. and Holmquist, L.E. (2011) 'Performing a check-in: emerging practices, norms and "conflicts" in location-sharing using foursquare'. *13th International Conference on Human Computer Interaction with Mobile Devices and Services (MobileHCI '11)*. New York: ACM, pp. 57–66.

Daniels, P.W. (1975) *Office Location*. London: Bell and Sons.

Forlano, L. (2013) 'Making waves: Urban Technology and the co-production of place'. *First Monday*, 18 (11).

Foursquare (2011) *10 Million*. https://foursquare.com/10million (Accessed: 1 October 2012).

Foursquare (2014a). www.foursquare.com (Accessed: 14 August 2014).

Foursquare (2014b) 'Our crowd-sourced places database has over 60,000,000 entries and 5,000,000,000 check-ins, and one major new partner – Microsoft', in *The Foursquare Blog*. http://blog.foursquare.com/post/75603461066/our-crowd-sourced-places-database-has-over: 2014 (Accessed: 14 August 2014).

Frith, J. (2012) *Constructing Location, One Check-in at a Time: Examining the Practices of Foursquare Users*. NC State University.

Gehl, J. (1987) *Life Between Buildings: Using Public Space*. New York: Van Nostrand Reinhold.

Gergen, K.J. (2002) 'The challenge of absent presence', in Katz, J.E. and Aakhus, M. (eds), *Perpetual Contact: Mobile Communication, Private Talk, Public Performance*. Cambridge: Cambridge University Press, pp. 227–241.

Goffman, E. (1963) *Behaviour in Public Places: Notes on the Social Organization of Gatherings*. New York: Free Press.

Goffman, E. (1983) 'The Interaction Order'. *American Sociological Review*, 48 (1), pp. 1–17.

Google (2012) *Our Mobile Planet: Global Smartphone Users*. http://goo.gl/WAFg7: Google.

Gordon, E. and de Souza e Silva, A. (2011) *Net Locality: Why Location Matters in a Networked World*. Chichester, UK: Wiley-Blackwell.

Gottmann, J. (1977) 'Megalopolis and Antipolis: The Telephone and the Structure of the City', in de Sola Pool, I. (ed.), *The Social Impact of the Telephone*. Cambridge, MA: The MIT Press, pp. 303–317.

GPS Bites (2012) 'The Top 10 List of Worst GPS Disasters and Sat Nav Mistakes'. *GPS Bites*. http://www.gpsbites.com/top-10-list-of-worst-gps-disasters-and-sat-nav-mistakes (Accessed: 1 October 2014).

Graham, S. (1997) 'Cities in the real-time age: the paradigm challenge of telecommunications to the conception and planning of urban space'. *Environment and Planning A*, 29 (1), pp. 105–127.

Graham, S. and Marvin, S. (1996) *Telecommunications and the City*. London: Routledge.

Greenfield, A. (2006) *Everyware: The Dawning Age of Ubiquitous Computing*. Berkeley, CA: New Riders.

Habuchi, I. (2005) 'Accelerating Reflexivity', in Ito, M., Okabe, D. and Matsuda, M. (eds), *Personal, Portable, Pedestrian: Mobile Phones in Japanese Life*. Cambridge, MA: The MIT Press.

Hall, E. (1966) *The Hidden Dimension*. Garden City, NY: Doubleday Anchor Books.

Hampton, K. (2007) 'Neighborhoods in the Network Society: The e-Neighbors Study'. *Information, Communication & Society*, 10 (5), pp. 714–748.

Hampton, K., Livio, O. and Sessions, L. (2010) 'The Social Life of Wireless Urban Spaces: Internet Use, Social Networks, and the Public Realm'. *Journal of Communication*, 60 (4), pp. 701–722.

Harrison, S. and Dourish, P. (1996) 'Re-Place-ing Space: The Roles of Space and Place in Collaborative Systems'. *ACM Conference Computer-Supported Cooperative Work CSCW'96* Boston, MA: ACM, pp. 67–76.

Harrison, T. and Stephen, T. (1999) 'Researching and Creating Community Networks', in Jones, S. (ed.), *Doing Internet Research: Critical Issues and Methods for Examining the Net* London: Sage Publications.

Hight, J. (2013) 'Narrative archealogy', in Buschauer, R. and Willis, K.S. (eds),*Locative Media: Multidisciplinary Perspectives on Media and Locality*. Bielefeld: Transcript Verlag.

Horan, T. (2000) *Digital Places: Building Our City of Bits*. Washington, DC: Urban Land Institute.

Humphreys, L. and Liao, T. (2013) 'Foursquare and the Parochialization of Public Space'. *First Monday*, 18 (11).

Hutchins, E. (1996) *Cognition in the Wild*. Cambridge, MA: The MIT Press.

Hyman, I., Boss, M., Wise, B., McKenzie, K. and Caggiano, J. (2010) 'Did you see the unicycling clown? Inattentional blindness while walking and talking on a cell phone'. *Applied Cognitive Psychology*, 24, pp. 597–607.

Ito, M. and Okabe, D. (2005) 'Technosocial situations: emergent structuring of mobile e-mail use', in Ito, M., Okabe, D. and Matsuda, M. (eds), *Personal, Portable, Pedestrian: Mobile Phones in Japanese Life*. Cambridge, MA: The MIT Press.

Ito, M., Okabe, D. and Andersen, K. (2010) 'Portable Objects in Three Global Cities: The Personalization of Urban Places', in Ling, R. and Campbell, S.W. (eds), *The Reconstruction of Space and Time: Mobile Communication Practices*. New Brunswick, NJ: Transaction Publishers, pp. 67–87.

Jacobs, J. (2002) *The Death and Life of Great American Cities*. New York: Random House.

Lawson, B. (2005) *The Language of Space*. Oxford: Architectural Press.

Lefebvre, H. (1991) *The Production of Space*. Oxford and Cambridge, MA: Blackwell.

Ling, R. and Yttri, B. (2002) 'Hyper-coordination via mobile phones in Norway', in Katz, J. and Aakhus, M. (eds), *Perpetual Contact: Mobile Communication, Private Talk, Public Performance*. Cambridge: Cambridge University Press, pp. 139–169.

Lofland, L.H. (1998) *The Public Realm: Exploring the City's Quintessential Social Territory*. Hawthorne, NY: Aldine de Gruyter.

Lynch, K. (1967) *The Image of the City*. Cambridge, MA: The MIT Press.

Manzo, L. and Perkins, D. (2006) 'Finding Common Ground: The Importance of Place Attachment to Community Participation and Planning'. *Journal of Planning Literature*, 20 pp. 335–350.

Massey, D. (2005) *For Space*. London: Sage.

McCullough, M. (2004) *Digital Ground: Architecture, Pervasive Computing and Environmental Knowing*. Cambridge, MA: The MIT Press.

Meyrowitz, J. (1985) *No Sense of Place: The Impact of Electronic Media on Social Behavior*. New York: Oxford University Press.

Mitchell, W.J. (1995) *City of Bits*. Cambridge, MA: The MIT Press.

Muzychenko, A. and Kats, P. (2015) *Most popular Foursquare venues*. Foursquare Stats. http://www.4sqstat.com/ (Accessed: 1 January 2015).

Needleman, R. (2009) 'Starbucks: Stay as long as you want'. *CNet*. http://www.cnet.com/uk/news/starbucks-stay-as-long-as-you-want/ (Accessed: 1 October 2014).

Oldenburg, R. (1999) *The Great Good Place: Cafés, Coffee Shops, Bookstores, Bars, Hair Salons, and Other Hangouts at the Heart of a Community*. New York: Marlowe & Company.

Pew Internet Project (2014) 'Social Networking Fact Sheet'. http://www.pewinternet.org/fact-sheets/social-networking-fact-sheet/.

Popper, B. and Hamburger, E. (2014). 'Meet Swarm: Foursquare's ambitious plan to split its app in two: To take on Yelp, Foursquare is moving beyond the check-in.' *The Verge*. http://www.theverge.com/2014/5/1/5666062/foursquare-swarm-new-app (Accessed: 1 October 2014).

Proshansky, H., Ittelson, W. and Rivlin, L. (eds) (1970) *Environmental Psychology: Man and His Physical Setting*. New York: Holt, Rinehart & Winston.

Relph, E. (1984) *Place and Placelessness*. London: Pion Limited.

Rheingold, H. (2000) *The Virtual Community: Homesteading on the Electronic Frontier* Cambridge, MA: The MIT Press.

Rogin, J. (2011, August 16). 'Meet the real mayor of the White House on Foursquare.' Foreign Policy (Retrieved: 2 August 2014).

Rowan, D. (2010) 'Revealed: the real mayor of London.' *Wired*. http://www.wired.co.uk/news/archive/2010-09/07/revealed-the-real-mayor-of-london (Accessed: 23 August 2014).

Scannell, P. (1996) *Radio, Television and Modern Life: A Phenomenological Approach*. Oxford: Blackwell.

Schegloff, E. (2002) 'Beginnings in the telephone', in Katz, J.E. and Aakhus, M. (eds), *Perpetual Contact: Mobile Communication, Private Talk, Public Performance*. Cambridge: Cambridge University Press, pp. 284–300.

Schwartz, R. (2014) 'Online Place Attachment: Exploring Technological Ties to Physical Places', in de Souza e Silva, A. and Sheller, M. (eds), *Mobility and Locative Media: Mobile Communication in Hybrid Spaces*. Abingdon: Routledge, pp. 85–100.

Tallentyre, M. (2013) 'Failed sat nav blamed for Chester-le-Street cyclist's A68 death'. *Northern Echo*.

Tate, R. (2014). 'How Foursquare Plans to Break Its App Without Enraging Users.' *Wired*. http://www.wired.com/2014/05/foursquare-swarm/ (Accessed: 1 October 2014).

Tuan, Y.-F. (1977) *Space and Place: The Perspective of Experience*. Minneapolis: University of Minnesota Press.

Tuters, M. (2004) 'The Locative Commons: Situating Location-Based Media in Urban Public Space'. *Futuresonic 2004*. Manchester, UK.

Undergleider, N. (2012) 'Bloomberg On Mayors Vs. Foursquare Mayors'. *Fast Company*. http://www.fastcompany.com/1826520/bloomberg-mayors-vs-foursquare-mayors (Accessed: 1 October 2014).

Weiser, M. (1991) 'The Computer for the 21st Century'. *Scientific American*, 265 (3), pp. 94–104.

Wellman, B. (2001a) 'Little Boxes, Glocalization, and Networked Individualism', in Tanabe, M., van den Besselaar, P. and Ishida, T. (eds), *Second Kyoto Workshop on Digital Cities II, Computational and Sociological Approaches*. London: Springer-Verlag, pp. 10–25.

Wellman, B. (2001b) 'Physical Place and Cyberplace: The Rise of Personalized Networking'. *International Journal of Urban and Regional Research*, 25 (2), pp. 227–252.

Whyte, W. (1988) *City: Rediscovering the Center*. New York: Doubleday.

Whyte, W.H. (1980) *The Social Life of Small Urban Space*. Washington, DC: The Conservation Foundation.

Willis, K.S. (2008) 'Places, Situations and Connections', in Aurigi, A. and Cindio, F.D. (eds), *Augmented Urban Spaces: Articulating the Physical and Electronic City*. Aldershot: Ashgate, pp. 9–26.

Willis, K.S. (2012) 'Being in Two Places at Once', in Abend, P., Haupts, T. and Müller, C. (eds), *Medialität der Nähe: Situationen – Praktiken – Diskurse*. Bielefeld: Transcript.

Zheng, Y. (2011) 'Location-Based Social Networks: Users', in Zheng, Y. and Zhou, X. (eds), *Computing with Spatial Trajectories*. New York: Springer.

3

Boundaries

3.1   DIGITAL THRESHOLDS

Café

3.1   My laptop
(with Wi-Fi dongle)
on outside table
at Morgenland
café, Berlin, 2006.

I spent quite a lot of time between 2006 and 2008 looking for Wi-Fi. Although it seemed to be pretty common in the city, finding it wasn't that easy. For me, it involved a modified practice known as 'war-walking', a form of digital divining, which literally meant systematically walking through an urban space with a laptop equipped with a Wi-Fi aerial, a GPS mouse and running a program that would detect Network ID's. The program would ping satisfyingly every time I picked up a new ID, and I found that as I entered certain types of urban space the pinging became almost constant. I was particularly interested in the distinction between open or public Wi-Fi nodes and closed or secure nodes, and whether there was any logic to the corresponding spatial boundaries of public and private space. Having done this in a series of European cities, I did start to get a sense of where public Wi-Fi nodes might be located. Interestingly it turned out that open Wi-Fi was often to be found in semi-public buildings or those with an 'open-to-all' atmosphere; public access seemed mostly to work on both a technological and social level.

During my Wi-Fi investigations and whilst visiting Berlin, I stumbled across café Morgenland in the Kreuzberg neighbourhood of Berlin, Germany, that advertised itself as having Wi-Fi (it was then called WLAN) which was useful as I needed to send an email. I took a seat on one of the wooden benches outside on the pavement, and opened up my rickety white plastic iBook. Back then I needed an incongruous-looking WLAN dongle to get a network connection, and I duly found and connected to the café Access Point that was accessible without a password. I sat for about an hour checking emails and although I wasn't thirsty I felt I should contribute something for using their Wi-Fi so then, for the first time since my arrival over an hour ago I went inside the café and dutifully bought a coffee. No one bothered me at my seat out in table on the street, and I seemed to be the only one working; it was midday and most people were busy finishing off their lunches. Café Morgenland is not one of Berlin's mega internet cafés such as St. Oberholz, on Rosenthaler Platz in the old East of the city, where rows of Apple logos shine out from the laptops of huddled, digital nomads. Café Mogenland prides itself on both its food and its good location; the advertising on its website is as follows; 'with the best view of the metro U1 we are a popular and world-renowned café and restaurant for young and old. You find us in the middle of Berlin – Kreuzberg near Görlitzer Bahnhof. We are one of the first Mediterranean restaurant here and very famous all over the world' (*Morgenland*, 2014). The café was a range of spaces; a building made of bricks and mortar, a place for eating food, a place to meet with others, a place that provides Wi-Fi, and also a website that provides information about the café. All of these spaces operated mutually independently; but the boundaries of where one ended and the other began were less clear. The café's tables and chairs also spilled out into the street, so that the physical edge of the space was also slightly ambiguous. But for me the café was a temporary workspace, and I realised that I was in a place that provided Wi-Fi first and was a café second. I probably could even have got away without buying the token coffee. In which case the functionality and the boundaries of the café as a place had slipped; it was a workplace where I communicated with people in remote locations through a network connection, now-and-then shifting or flicking back into a place to eat and drink or take in the street atmosphere. This seems pretty unremarkable now, where Wi-Fi-enabled cafés are increasingly used as transitory work places, but less than ten years ago it was unfamiliar territory.

Spatial thresholds

Something as simple as opening or closing a Wi-Fi connection allows us to flick between different places. Yet the threshold in between these places is less clear. Going 'online' involves a transition of a click to access information, whereas in a building you enter through a door or look through a window. In the built space, the transition is mediated by a threshold, which is 'a point where the boundary between inside and outside can be opened; space loosens up, and a wide range of perceptions, movements, and social encounters become possible' (Stevens, 2007, p. 73). The threshold is the point where two different worlds meet. Doors, corridors, windows, turnstiles, porches, terraces and stairways distinguish inside from outside and mediate people's passage between them. The threshold according to Bourdieu 'is the site of a meeting of contraries as well as of a logical inversion and … as the necessary meeting and crossing point between the two spaces, defined in terms of socially qualified body movements, it is the place where the world is reversed' (1992, pp. 281–282).

It also sets up rules and frameworks for social behaviour, and has an inherent social logic, so that a threshold is also a restricted space; its design always constrains people's behaviour and their perceptions (Hillier and Hanson, 1984). Locations and the settings that constitute them have rules associated with them about appropriate behaviour, who belongs there and has (how much) access, who controls or uses resources (and which). Rapoport (1994), in an anthropological reading of the threshold, points out that behaviour and activity systems are organised relative to spatial organisation and built environments, as these are expressed in systems of settings. One result of this process is that boundaries are created, and then maintained or controlled. Rapoport points out that 'boundaries are selectively permeable; various people are admitted to or excluded from various spatial domains or settings and may penetrate deeply or just minimally, may become central or controlling or remain peripheral, depending on who they are and what rules apply' (1994, p. 82). But as seen from the café example above, technologies such as Wi-Fi cause us to reassess and reconfigure spatial frameworks framed around spatial proximity (i.e. distance) and bounded-ness.

Edges and transitions

The predominant visuo-spatial way of understanding urban spaces is affected by technologies such as Wi-Fi since they have very little visual presence. Territories in layered media spaces are not solely defined by the physical properties of 'real world' objects and spaces, but also extend to include the specific ranges of technologies. For instance RFID enables interaction within a radius of approximately two metres, whereas Wi-Fi nodes offer access within a range of up to one hundred meters. Connection and separation are less defined by the physical or embodied proximity and more by the availability for engagement. For example, a wall which traditionally defines the degree of visual and physical separation between one space and another does not act as a barrier for a Wi-Fi signal, as anyone who has ever accessed the internet through a neighbour's Wi-Fi connection will know. In this way separation is less defined by material barriers, and more through the range of the particular technology for including or excluding a situation. These changed patterns of separation create new ways of realising bounded spaces.

This is the case with Wi-Fi; a 'wireless' technology that works on emitting waves of a particular frequency range (2.4Ghz). However Wi-Fi is also a technology that has an infrastructure that is fixed in space; Wi-Fi access has a range that may extend spatially, but it is structured around a series of nodes which are literally black boxes emitting a wireless signal that have a specific location in space. Wi-Fi waves pass through the visual and material boundaries of walls and windows, but are blocked by metallic materials. So, concrete alone is no problem, but if it has a steel reinforcing mesh, this will act as a barrier. It also needs a line of sight to work, and the strength of the signal depreciates over distance. The boundaries of Wi-Fi have some similarities to physical boundaries, but since they are waves and not topological or material relations the two do not map easily on to one another. Access to a Wi-Fi node requires that you know the name of the node, have a device, a piece of software and password or key to gain entry, but the transition is instantaneous. Walking in to a building requires some physical motion and a negotiation of access, but the key to Wi-Fi and the key to the door are not the same thing. If I enter a café on foot, I do not automatically negotiate Wi-Fi access. Conversely, the door of the café may be closed, but I may still be able to access the spilling range of the Wi-Fi node out in the public space of the street. According to Mitchell 'where walls once established relatively clear and stable boundaries among social settings, mobile devices create unexpected and sometimes difficult to manage juxtapositions' (2004, p. 81). Walls, doors and other points of negotiation of territory and access do not have the same defining authority when Wi-Fi and other technologies create other topological thresholds and boundaries. The challenge is to understand what this means for the way we act and design our future spaces, and how we make sense of existing ones.

Just connect

'With, between, in, before, far' according to Mackenzie, in his study on wirelessness, are the relations we encounter incessantly in a wireless world. In fact 'life is lived far more in these relations than in the disjunctive relations associated with things and entities' (Mackenzie, 2010, p. 20). Concepts of proximity (Hall, 1966), threshold (Stevens, 2007) and region (Lynch, 1967) are brought into question by readings of technological change; leading to terms such as transition space (Gottdiener, 2001), code/space (Kitchin and Dodge, 2011) and wirelessness (Mackenzie, 2010). In the interaction with technology, regions are not only defined by spatial extents, but also by patterns of informational or social access and structures. These new spatial frameworks work on the basis that 'connection is more important that division' (Varnelis, 2012, p. 146); where 'networked publics' (Ito and Okabe, 2005; Varnelis, 2012) challenge the idea of a binary separation between public and private space and behaviour. Boundaries are still an omnipresent characteristic of space, but moving in and out of bounded zones can occur much like the flicking of a switch, rather than involving some form of graduated change. According to Graham and Marvin 'the legacy of the physical and locational approach, and the continued attachment to independent notions of time and space, still severely limits the degree to which telecommunications-based effects can genuinely be incorporated into many approaches to urban analysis and policy-making' (1996, p. 73). The limits in exploring the potentials and working with the problems of fluid boundaries are complicated by the fact that we lack the conceptual tools to make sense of the changes that are occurring. In the next section I attempt to unpack some of these theoretical constructs and assess them anew.

Aims

This chapter aims to explore how boundaries; that is structures of spatial separation and constraints are understood. It starts out be exploring how some of the binaries of our built world; public/private, inside/outside and open and closed, are affected by patterns of informational or social access. Socially constructed concepts of territory, boundary and region are initially investigated in terms of how they are negotiated through patterns of inclusion and exclusion. The properties of wirelessness and wireless connections are also explored to give context to how these networked infrastructures operate. The chapter uses the lens of work and the changing boundaries between living and working in a historical sense to understand the historical context. I then focus in on a range of changing conditions and explore these in more detail. The changing nature of boundaries is highlighted in how digital interactions are resulting in different models of private and public, with a 'public-by-default' replacing the traditional model of choosing to be 'in public'. Similarly, the uses of technology and social networks create a new model of being 'together', and models of shared space that are responsive and networked. The challenges of those excluded, or not given access to the digital world is shown to be a significant and real issue and that whilst some boundaries may break down, conversely new boundaries of exclusion emerge. These conditions are tested out through a case study of examples of public Wi-Fi networks that have been installed in urban spaces. The way that these networks affect patterns of access, public and private territories and also how they create digital divides is discussed. Finally the chapter explores the implications of these findings for the field of architecture, and in particular focuses on the challenge of the relationship between code and space, as well as the opportunities for different models of shared space that these new configurations open up.

3.2   DIGITAL BOUNDARIES

Spatial and digital separation

In his essay, 'Bridge and Door', Simmel (1994 [1909]) distinguishes between the ways we connect and separate spaces. He points out that in the 'the immediate as well as the symbolic sense, in the physical as well as the intellectual sense, we are at any moment those who separate the connected or connect the separate' (1994 ([1909], p. 171). For Simmel, the bridge expresses in a tangible, object-form the 'will to connection' (1994 [1909], p. 6) that overcomes the physical separation of the two banks of the river. In contrast the door captures a duality between closed and open, which frames the flows between the inside and the outside:

'[t]hus the door becomes the image of the boundary point' (Simmel, 1994 [1909], p. 7) so that, 'in the unity, the bounded and the boundaryless adjoin one another, not in the dead geometric form of a mere separating wall, but rather as the possibility of a permanent interchange' (Simmel, 1994 [1909], pp. 7–8). Benjamin (2002), in his Arcades project, makes a conceptual distinction between the boundary and threshold; 'the threshold must be carefully distinguished from the boundary. A threshold (schwelle) is a zone. Transformation, passage, wave action. Whereas a boundary is a line that separates, a threshold is 'a zone of transition'. Thresholds are interesting since they allow passages across them, transitions between spheres or states; while boundaries tend to halt movements, thresholds invite innovative change' (2002, p. 494). Separation is also managed by material and social norms and rules; physical barriers and locks provide the most obvious controls on the use of space, but an individual's behaviour is also constrained by what they think is appropriate, admissible or possible. Networks of mobility, globalisation and technology destabilise the material conditions of territory and boundaries. Law links this with a process of dematerialisation which has effects that are either 'durable or otherwise as a function of its location in the networks of the social' (1994, p. 102). On the one hand, traditional boundaries, whether political or spatial (or both), are losing some of their material capacity to separate. For example, just look at how multinationals such as Google and Amazon evade national tax structures by claiming statelessness (Drucker, 2013). On the other physical structures used to separate such as walls, fences and ledges, which are often supposed to delimit space and behaviour, 'can be sat upon climbed onto and used to display banners or items for sale, their looseness is a product of affordances which such boundaries provide' (Gibson, 1979). On a social level, the physical qualities of urban space can frame opportunities for expression and social engagement. Increasingly we use the world around us not to divide or create difference, but to connect, and this often means moving beyond the use of spatial metaphors to define separation. Sites are defined not by spatial boundaries or scales, but by types and lines of activity, and spaces emerge through the networks connecting the sites (Latour, 2005). One of the characteristics of networked technologies is that they are enabled by, and operate in a state of 'wirelessness' (Mackenzie, 2010). This highlights the inherent immateriality of many connections we make with technology; Wi-Fi, GPS, Mobile Phones and RFID all work on frequencies that use the air to propagate. Hence, wirelessness 'as a contemporary mode of experience concerns a topologically problematic space' (Mackenzie, 2010, p. 94). Shepard highlights that 'what is significant is that as these mobile devices become ubiquitous in urban environments (and in many places they already are), the technicity of architecture as the primary technology of space-making is challenged by the spatial transductions these devices afford' (Shepard, 2011, p. 25). So, devices we carry with us, and the way we interact with them start to reframe the physical properties of places. Fujimoto terms these 'territory machines' in that they are 'capable of transforming any space – a subway train seat, a grocery store aisle, a street corner – into one's own room and personal paradise' (2005,

p. 77). Shepard describes how an iPod 'becomes a tool for organising space, time and the boundaries around the body in public space' (2011, p. 25). People set up personal territories in their fluid, mobile lives. For example Rayner (1998) recounts the practices of a businessman at an airport where 'the first thing that most of them do, before opening their laptops or helping themselves to coffee, sometimes even before taking off their coats and dumping their packs, is to reach for a phone and punch in a calling card number. ... It's a check in, not with the airline or even the office, but with themselves'. In fact, airports have started to install charging points in their lounges (Heathrow Airport), that create micro-territories of plugged-in travellers. The technological characteristics (both positive and negative – either a network 'black hole' or a Wi-Fi public hotspot) create an intense focus either on the physical space or the media space; these are infrastructural and institutional 'moorings' or places that configure and enable mobilities (Hannam, Sheller and Urry, 2006). In a study of mobile practices in three European cities Lasen observed the prevalence of mobile access at the exits of Underground (metro) stations that resulted in the formation of 'open-air wireless phone booths' (2003). As we increasingly focus on door-like structures to connect, rather than mental and physical bridges that separate, the spatial characteristics of the environment no longer work in isolation. Instead the combination of the physical characteristics and the technological characteristics create territories that frame behaviour and the way we inhabit space.

Public/private

The distinction between public and private has been a key spatial framework for built space. Jacobs treated it as one of defining characteristics of an urban street where 'there must be a clear demarcation between what is public space and what is private space. Public and private spaces cannot ooze into each other as they do typically in suburban settings or in projects' (2002, p. 35). Distinctions between private and public are literally quantified by ownership, but are socially enacted as patterns of inclusion and exclusion. Walls and doors have typically been the frameworks that define whether someone is allowed to enter a region such as room, or is excluded from it. According to Goffman a person 'will often be required to show some regard for the physical boundary around it, when there is one'. But walls and doors are also replaceable since 'the work walls do, they do in part because they are honoured or socially recognised as communication barriers, giving rise ... to the possibility of conventional situational closure; in the absence of actual physical closure' (1963, pp. 151–152). Hall (1966), who introduced the study of proxemics, makes a distinction between public space and personal space of the body based on a culturally constructed relationship in metric space. Hall categorises 'proxemic' features as either fixed, semi-fixed or dynamic (1966). Fixed-feature spaces are formed by walls and territorial boundaries; semi-fixed features are spatial constellations formed by mobile elements such as curtains, screens, movable partitions, and furniture arrangement. In the use of communication devices such as mobile phones, there is a recognisable shift from proximity that is based on measurable distances to that where the media practice itself establishes degrees of social closeness (Willis, 2012). Thus, a mobile phone call may create an intimate space among those communicating, but they may be physically very far from each other, and similarly the presence of a stranger can be displayed as very close to someone in space on a platform such as Foursquare but the person will interact with them as though they were a stranger and physically distant from them.

This causes a reflection of what constitutes privacy, something that has been a topic of much debate, often as the result of the everyday implications of using social networking sites such as Facebook. Boyd terms these 'networked publics' and highlights how 'the blurring of public and private: without control over context, public and private become meaningless binaries, are scaled in new ways, and are difficult to maintain as distinct' (2010, p. 10). She points out that 'we've moved from a world that is "private-by-default, public-through-effort" to one that is 'public-by-default, private-with-effort' (boyd, 2014). In many ways social networking has resulted in a dissolving of the default of a private social space, to a default of that which is displayed in public. But, in parallel with this, the use of social media platforms such as Snapchat and texting create new kinds of selective private space or 'telecocoons' (Habuchi, 2005, p. 178). In particular the rise in ownership of mobile phones creates new kinds of bounded places, which has been found to be particularly valuable to Japanese teenagers who, it has been suggested, lack private physical places (Ito and Okabe, 2005, p. 260). We cannot rely on the fixed features of a wall to define a private space or a door to provide the transition between public and private; wireless communication is reframing social constructions of personal and public space. But, just as private spaces are opening up to public view, so public spaces are being occupied and transformed into private cocoons of communication.

Rules, norms and code space

Goffman (1963) introduced how the rules and norms of social interaction, although invisible, could be measurable; creating foregrounds and backgrounds, actors and audience. Just as invisibly, the code and rules of technological infrastructure, software and devices 'are installing a new kind of automatically reproduced background and whose nature is only now starting to become clear' (Thrift and French, 2002, p. 309). Kitchin and Dodge (2011) argue that the codes of technological infrastructure that are intimately woven into our everyday lives create a 'code/space' (Kitchin and Dodge, 2011). Dodge and Kitchin equate code/space with the mundane corridors and lounges of air travel, and highlights how 'the vast bulk of time in code/space either in the airport or in the air is largely banal: spaces of chat, gossip, waiting, fidgeting, reading, staring, eating, and so on' (2004, p. 204). These transition spaces may equate with connections in the infrastructure, but there is a more pervasive re-structuring of spatial relations. In Shepard's view of the 'sentient city' this restructuring extends beyond airports and out into every space of the city. This happens when computing capacity is embedded within and distributed through the material fabric of everyday urban life, so that 'on any day we pass through transportation systems using magnetic strip or RFID tags to pay a fare; we coordinate meeting times and places through SMS text messaging on the run; we cluster in cafés and parks where Wi-Fi is free; we move in and out of spaces blanketed with CCTV surveillance systems monitored by computer vision systems. Artefacts and systems we interact with daily collect, store and process information about us, or are activated by our movements and transactions' (Shepard, 2011, pp. 19–20). The key issue with this re-ordering of the rules of code onto the physical and spatial rules is what happens when the code/space fails. For example, an office often ceases to function if the broadband connection goes down, even if it still has power. In a code/space 'the domination of code is so pervasive that if one half of the dyad is put out of action then the entire code/space fails' (Dodge and Kitchen, 2004, p. 198). This suggests that, although the rules and frameworks or network infrastructures are embodied through the performances and interactions of the people within the space, this is not a fully reciprocal or 'sticky' relationship.

3.3   HISTORICAL CONTEXT

Technological boundaries: from wired to wirelessness

In 1902, at the turn of the twentieth century, the laying of the first telegraph cable across the Pacific was completed, finally encircling the world. It enabled the connection of remote places and the transmission of messages from one remote location to another. In a parallel development, between 1886 and 1868, a physicist named Hertz, demonstrated that the transmission of electricity was actually through waves; electro-magnetic radiation. His work preceded the invention of the radio, a technology that no longer needed a physical connection, a wire, between two points to enable communication. This was an era which heralded an industrialised society, where communication started to be broadcast or available to a mass, rather than between individuals. The invention of the radio, exploited by Marconi, marked the advent of 'wirelessness' where a message transmitted from one point could be received at any number of points, and all receivers would have the generally same experience. At the time, this caused concern because the fact that it was not possible to know or decide who would receive the message was perceived as a fundamental weakness of the transmission process (Winston, 1998, p. 76). By the 1920s the transmission of wireless media meant that the boundaries of the communication process thus became transformed; from a reciprocal exchange between two defined points, to a centralised transmitter with multiple receivers that could be located anywhere within a radius of the transmitted waves. The radio impacted on other boundaries, such as those between the city and rural regions, and between working life and home life. For the former, access to broadcasts for the rural population, such as farmers, in the first decades of the twentieth century meant that geographical isolation did not necessarily equate with social or cultural isolation. According to a 1922 article in *The New York Times*: '[radio] not only can connect every farm with the nearest city but with the entire world' (n.n. 1922). Although a wired technology, the telephone had a similar impact on the spatial organisation of both urban and rural spaces. De Sola Pool makes a compelling case for how the telephone reorganised boundaries in cities in the US, and in particular resulted in the growth of the suburbs and the intensification of urban centres. De Sola Pool acknowledges the role of the car as a factor in this process; 'the automobile and the telephone help make it possible for metropolitan regions to spread over thousands of square miles … it helped dissolve the solid knots of traditional business neighbourhoods, but at a later stage it helped disperse those downtowns to new suburban business and shopping centers' (de Sola Pool et al., 1977, pp. 141–142). De Sola Pool also points out that the commercial development of the telephone supported the development of zoning and the formation of defined neighbourhoods and the definition of city boundaries, in order to make the delivery of the service more economically efficient and profitable. Thus, although the potential of the technology, in the first half of the twentieth century, was to create spatial homogenisation and dispersal, the initial infrastructural requirements drove a rationalisation and definition of spatial boundaries. It was not until 1970s and the introduction of mobile technologies that wirelessness enabled a more complete dissolving of spatial territory in relation to wireless access. The mobile phone worked on a new infrastructural framework; the use of 'cells' much like a honeycomb structure, where each cell was served by a mast or transmitter. The ability of the communication connection to follow the mobile user, created more fluid topologies; ones that are carried by people as they move around, rather than inscribed in the place. Bounded edges are created only at the point where there are breaks in connectivity; so a 'not-spot' or rural location with no signal becomes territorially very different even if the topography and visual characteristics do not change from its immediate surroundings. The introduction of Wi-Fi had similar characteristics; whilst within range of a Wi-Fi node connectivity equates to a territory, as you move out of range a different kind of place is entered. Yet the provision of Wi-Fi works on a different scale; it operates on connection to individual nodes, and moving between these nodes is less seamless. Wi-Fi zones tend to be linked to places; for example Starbucks and other café's offer branded Wi-Fi available within their premises. Some cities, companies and educational institutions create Wi-Fi zones comprising a mesh of nodes; yet even here moving from one space to another results in losing a connection and needing to reconnect, creating a boundary between one space and the next.

Social boundaries: from private enclosures to networked publics

In architectural space, one of the fundamental ways that boundaries are understood is the distinction between inside and outside, and this encompasses the act of differentiating spaces according to whether they are public or private. It is traced back to the beginning of the seventeenth century when the private house started to develop as a broadly popular type (Riley, 1999, p. 10). The boundary between public and private is an important foundation of social behaviour; public places have different rules of behaviour and access to private spaces. In the 1970s, Altman introduced the idea that private space defined certain types of behaviour and that 'privacy is an interpersonal boundary-control process, which paces and regulates interaction with other' (1975, p. 10). In so doing 'each individual is continually engaged in a personal adjustment process in which he balances the desire for privacy with the desire for disclosure and communication of himself to others' (Westin, 1967, p. 7). Also in the 1970s the sociologist Goffman (1990) introduced the concept that people construct territories based on access to situations, and outlined an approach to explain how unspoken rules defined behaviour in situational encounters and thus how places had socially constructed boundaries. Ten years later, Meyrowitz, in a review of Goffman's work, argued that the consequence of communications technologies in urban settings is that multiple social realities can occur in one place. The same physical space may be caught within the domain of two different social occasions. Meyrowitz, states that the consequence of communication technologies for social interaction is that it is 'no longer the physical setting itself that determines the nature of the interaction, but the patterns of social information flow' (1985, pp. 36–37). These technologies enabled a disassociation to occur between the setting and social behaviour. Further it meant that media affect traditional boundaries between private and public. A number of authors have argued that the growth in mobile phone and internet use has meant a privatisation or parochialisation of public space (Humphreys and Liao, 2013; Ling, 1997). But in parallel the rise of networked publics has been documented, where new publics emerge, as a result of media behaviours, that transcend spatial boundaries (Ito, 2012; Varnelis, 2012; Gordon and de Souza e Silva, 2011). Mackenzie argues that these conditions of wirelessness means developing alternatives to phenomenological, existential or socio-psychological accounts of experience since 'experience overflows the borders and boundaries that mark out the principal lived functions of subjectivity-self, institution, identity and difference, object, image and place' (2008, p. 15). Instead Mackenzie makes a case to move away from trying to define experience as a bounded entity and to treat it as a transition. This focuses, not on a Euclidean notion of spatial container with an inside and outside, but on an experience of a space that is constructed through transitions from one situation to another and how these connections are established and managed. It is a space experienced at the threshold between one situation and another.

3.4 THRESHOLDS

Filtering and the re-construction of privacy

As Mitchell points out 'the enclosing surfaces of the constituent spaces – walls, floors, ceilings and roofs – provide not only shelter, but privacy' (Mitchell, 1995, pp. 121–122). In the built world we manage a more differentiated condition of privacy through filtering devices such as curtains, shutters and glass walls. This gives us agency over how gradations of privacy are controlled; we open the curtains to let eyes in whereas glass walls that are opaque in the day become transparent at night. Boyd finds that in order to manage privacy online it requires that 'people have agency in their environment and that they are able to understand any given social situation so as to adjust how they present themselves and determine what information they share' (boyd, 2011). In a wireless world we have learnt that passwords and choosing who to 'friend' or 'like' are practices that work like curtains and glass walls; they let certain people in and exclude others (well, not quite as the criticism of Facebook and Co.'s privacy laws show (Wikipedia, n.n.), but that's we like to think). Humphreys has termed this 'filtering', and found that it also spills over into spatial settings. In a 2007 study of Dodgeball (a Foursquare predecessor) she found that people used the app to meet existing friends out on the town, and in meeting those friends did not necessarily connect to the general public leading to 'social molecuralisation' (Humphreys, 2007), that are similar to Wellman's 'little boxes' of specialised social networks (2001). The diversity and inclusiveness of urban spaces becomes masked by 'net localities', so that rather than chance encounters of difference, mobile social networks facilitate 'chance encounters of sameness' (Gordon and de Souza e Silva, 2011, 146). Gordan and de Souza e Silva found filtering effects in the way users 'customise' public spaces with the features of location-based networking apps, which allows users to choose the type of person to show on their map. This filtering or customisation of space creates a type of 'differential public space' in which physically co-located people experience things very differently (Gordon and de Souza e Silva, 2011, p. 146). Filtering is a practice that reclaims socially constructed territories in a 'public-by-default' world.

In a world that is public-by default (boyd, 2014) the onus has shifted to a proactive reconstruction of private spaces through a range of practices. In the working environment, this can be exemplified in the reaction to trend towards the 'open-plan' office. An early example of this is the attempt made by in 1993 by Chiat and Day, a New York based global advertising group, inspired by the promised of untethered ICTs, went 'public by default' in their office headquarters. In a complete re-organisation the company took away all desks and offices, in a move heralded by Time magazine who reported how employees 'thoroughly armed with the modern weaponry of the road warrior … the telecommuters of Chiat/Day are among the forerunners of employment in the information age' (Berger, 1999). Faced with an almost complete loss of territory and privacy the workers rebelled, complaining that 'in a virtual office, you can't hide' and by mid-1995 the more formal structures of a spatial office were returned (Berger, 1999). Ten years later 'public by default' in office spatial design is fairly common; unwired technology and new working practices means that work can and does happen anywhere.

A number of authors have reported on how the mobile phone is used to construct private territories (Lasen, 2003; Katz and Aakhus, 2002; Höflich, 2006; Humphreys and Liao, 2013). For instance, Pertierra (2005) explains how Filipinos, 'without a room of their own' use their mobile phones to create a private space. Whereas Höflich (2006), in a study that linked behaviour in an urban public square with mobile phone usage found that users of a public square walked in circles whilst engaged in a telephone call, oblivious to the natural linear flow of people moving through the space. Their movements constructed a territory, a personal space of their media interaction, which was independent from that of others in the social setting. In a public-by-default scenario one of the main acts of agency available to anyone is to switch off. For example, Barack Obama has access to a 'portable zone of secrecy' (Schmidt and Schmitt, 2013) that is literally a portable tent set up in whichever country he is in which has been designed to shield from all electronic signals. In order to set up the secure zone, Obama's staff spend time in advance of a visit 'locating and securing a certain area where the tent could be placed', and it has to be positioned carefully in relation to windows, and concentrations of people (Schmidt and Schmitt, 2013). Meanwhile in a publicity stunt in 2012, Nestlé KitKat®, known for their strapline 'Take a Break' created a response to the condition where 'people are constantly online' by installing a brightly painted red bench as a 'Free No-Wi-Fi Zone' in the city center. In a radius of five meters, the bench 'blocked all signals so people could escape e-mails, updates, tags or likes' (2012). One way to reclaim agency in a public-by-default society is to remove oneself from the public sphere through temporary structures, whether tents or benches or a quiet corner; each of them is about constructing a territory that shields from the waves of ubiquitous networks.

Peer-to-peer publics and shared spaces

If territory is traditionally shaped by ownership of property (Altman, 1975) one of the biggest challenges (and opportunities) of the next ten years is what is known as the 'sharing economy'. According to an *Economist* article, on one night in 2013, 40,000 people rented accommodation from a service that offers 250,000 rooms in 30,000 cities in 192 countries (n.n., 2013). The bookings were made online through a site named AirbNb, so named because the founders initially used airbeds to host their first guests on their living room floor. Similar sharing sites, such as Uber, Lyft and TaskRabbit extend the concept to car sharing and everyday tasks. Unlike earlier reciprocal platforms such as the bartering system LETS (Croft, n.d.), money does change hands, but the difference between renting is that the exchange is within a peer-to-peer network based on trust and reputation.

With over sixty locations worldwide, Impact Hubs comprise a network of co-working spaces aimed at social enterprise entrepreneurs. Described as 'an innovation lab, a business incubator and a social enterprise community centre' the hubs offer membership, in a model different to co-working or office sharing (Impact Hubs, 2014). Membership is not based on ownership or renting of the space, but is a subscription service where access is bought for chunks of time. Internally the spaces are designed around creating 'spaces for meaningful encounters': they are divided into a central room for flexible co-working and events, a series of semi-open meeting rooms, a secluded library for quiet thinking, and a community kitchen or café where people can eat together (Impact Hubs, 2014). The shared space model also extends to buildings that operate to share resources within a community. In USA there are over forty community tool libraries, with others in Canada, Sweden and Israel. Aimed at creating a resource of tools for local neighbourhoods, they also operate on a membership format and work by accepting tool donations from the community and then lending those tools out for free to anyone with ID. They also offer training sessions, such as how to fix everyday items. These hubs of the sharing economy disrupt territorial models of individual ownership and monetary exchange. The co-working hub and tool library are built spaces that do not represent fixed structures, but rather hosts for exchange and encounter that require some form of negotiation.

The street is also opening up to disrupt rule-based structures that create fixed spatial boundaries. Shared space, originated by Hans Monderman, is an approach to street urban design where all forms of control are removed. The philosophy is that absence of all of those features forces all users of the space to negotiate passage through the space via eye contact and person-to-person negotiation. According to Monderman 'people here have to find their own way, negotiate for themselves, use their own brains' (van de Vliet, 2013). A digital version of this has been tested by Audi and BIG, in a 190 m² three-dimensional LED installation where the public space is shared between pedestrians and driverless cars. In their prototype, the entire road surface is infused with a continuous flow of information allowing for real-time interaction between vehicles and their environment, whilst 3D cameras track the movement of passers-by processing the data into a generative artwork that feeds back into the LED panels. According to Andreas Klok, one of the designers; 'infusing the surface of the city with information technology will create a new kind of true shared space – a condition similar to a public square. In a single day, the function of the street may alternate multiple times between entirely pedestrian, vehicular, or even recreational functions' (Furuto, 2011). The technology-enabled version of shared space operates on different model in the sense that the LED control system re-introduces a code/space of a rule-based syntax into the interaction. This presents issues, such as what happens when unexpected situations occur in the interaction, or there is a technical failure. But the potential of shared spaces leads to an architecture whose forms have not been predetermined by the architect who sets the rules but is constantly being recomposed and renegotiated by the people using the space.

Access, inclusion and digital divides

Just as informational access can include, it can also exclude. Digital divides are inequalities between people 'in their access to, use of, or knowledge of information and communication technologies' (Warschauer, 2003, p. 1). Digital divides can be social; often experienced by older people or those with little access to education, who do not have access to computers in their everyday lives. Digital divides can also be linked to technical 'gaps' in the network, that are often more prevalent in rural areas. An April 2012 Pew Center study showed eighty-eight per cent of Americans over the age of eighteen have a cellphone (smartphone or otherwise), and sixty-six per cent of Americans aged eighteen to twenty-nine have smartphones. But only fifty-seven per cent of Americans have a laptop. The survey also stated, 'Among smartphone owners, young adults, minorities, those with no college experience and those with lower household income levels are more likely than other groups to say that their phone is their main source of Internet access' (Zickuhr and Smith, 2012). In the lowest income brackets, families don't have broadband at home, so public access is important. The availability of public wireless can help to connect individuals to the information society and knowledge economy, by reducing economic and geographic access barriers and thus break down boundaries.

3.5   CASE STUDY: PUBLIC WIRELESS NETWORKS

In order to test out how boundaries may be changing as the result of technology, in the next section I investigate public Wi-Fi. Wi-Fi is a wireless technology that allows an electronic device to exchange data or connect to the Internet, and is one of the most common ways we connect to the Internet, whether it is through a laptop or a smartphone at home or at work on the move (Figure 3.2). According to recent statistics; over seventy per cent of wireless traffic travels over Wi-Fi connections (Kelleher, 2013). The case study approach investigates a series of pioneering public access Wi-Fi and considers how they establish changing thresholds between public and private space and between work and home. I explore different models of Wi-Fi coverage, from single Wi-Fi nodes or access points through to large-scale urban Wi-Fi networks. This also opens up an exploration of the way that that differences between commercial, municipal and community Wi-Fi may help to reduce digital divides and contribute to the breaking down of social, economic and geographical boundaries.

3.2   Laptop user of free Wi-Fi in Bryant Park, New York (image courtesy of Bryant Park Corporation).

Home and work, private and public

Between 2002 and 2004 a number of initiatives developed to take advantage of the emergence of the growing bandwidth and robustness of Internet networks (this was before mobile internet on smartphones). These projects sought to create publicly accessible Wi-Fi, and so to offer public wireless Internet for social benefits. In 2002 in New York, the Wireless Internet Project set up an open wireless network within a public park setting in Bryant Park, New York, USA. The aim of the Bryant park wireless project was to provide free, unrestricted broadband Internet access to park visitors with the intention of changing public behaviour through the introduction of wireless availability in a park setting. In 2004, a similar initiative in UK, a well-organised, community-led mesh of public wireless nodes was established in Deptford, London, UK, called 'Boundless'. It developed out of the two initiatives that were the first to offer free Wi-Fi access in London; the Clink Street and the Consume network (Priest, 2004). The Boundless network was a community mesh network based on user-owned and operated local Internet access. In 2003, Île Sans Fil (which translates to 'Island Without Wires') in Montréal, Québec, Canada was established as a non-profit organisation operating a network of free wireless access points (or hotspots) in public locations. The project concentrated on the installation of wireless access points in publicly-accessible locations like libraries, bistros, cafés, parks, community centres and was aimed at mobile professionals, students and freelance workers. What all three projects; Bryant Park Wi-Fi, Boundless and Île Sans Fil, had in common was that they sought to harness publicly available Wi-Fi to reconfigure the boundaries of the physical and social spaces within which they were situated. Bryant Park sought to regenerate a run-down urban park to be an office; according to their publicity it would be possible to 'go wireless and turn Bryant Park into your new office. Your clients will be impressed with your front lobby' (Bryant Park Corporation, 2014). Boundless sought to develop and extend the existing public space in a deprived area of London, by adding a second publicly accessible Wi-Fi layer. Ile Saint Fils aimed to establish a public, collective space with the introduction of its network.

In Bryant Park, a public city park, the introduction of Wi-Fi to create an office in the park was successful with thousands of users per day logging on. But the consequence was that the openly public boundaries of the park became overlaid with private spaces of home and work. According to a study of Wi-Fi users in the park in 2006 and 2007 the primary purpose of the visit to the park was to use Wi-Fi for both work and personal use (sixty-three per cent), a smaller number of users reported using Wi-Fi for personal use only (twenty-eight per cent), and even fewer said that they use Wi-Fi only for work (eleven per cent) and Forlano (2009) concluded that 'people explicitly go to public spaces such as parks and cafés in order to do personal activities' (pp. 348–349). A study by Hampton of Wi-Fi use in Bryant Park in 2007 found that the behaviour of Wi-Fi users was equivalent to them creating a private space since they showed 'reduced attention to surroundings … and a focus on private, head-down activities' (Hampton, Livio and Sessions, 2010, p. 19). Interestingly Hampton found that although Wi-Fi users withdrew in the park space, the range of their online activities whilst in the park was much more public and sociable. This matches with Forlano's finding about the user's reports of how they mixed both work and personal use, rather than only work focussed Internet use. This creation of privately bounded spaces within the public space creates new spatial patterns of Wi-Fi use, through clustering of activity within specific areas (Hampton, Livio and Sessions, 2010, p. 24).

Interestingly the analysis of patterns of usage of the public Wi-Fi nodes in both Ile sans Fils and Boundless reveal that private users in their homes were also accessing the Internet located adjacent to the public Wi-Fi. Crow and Miller found the number of logins to Ile sans Fils dedicated portal (2006, p. 224) was far greater than those they could actually observe in the space (2006, p. 34), and concluded that 'many individuals using the ISF network were not sitting in cafés as the network designers intended, some individuals may be able to access an ISF hotspot signal in their own homes, rather than purchasing Internet service from an ISP' (Crow and Miller, 2006, p. 12). With the Boundless network this private usage could be seen both spatially and temporally. In the Wi-Fi mapping study I described at the introduction of this chapter I studied the link between the physical location of Wi-Fi nodes and the patterns of use in the Boundless network in Deptford, London. I found that the usage in public buildings as part of the Boundless networks (the Albany Theatre, the Laban Theatre and Creekside Community Wildlife Centre) suggests that the Wi-Fi nodes were not being accessed by actual visitors to the community centres. The Wi-Fi nodes in these buildings had periods of high usage after six o'clock pm and often during the middle of the night (Willis, 2006). These buildings were all closed at night time, so since all these buildings are located in the middle of residential tower blocks, it suggests residents in adjacent homes were accessing the Wi-Fi nodes. Not only was the public space of the Wi-Fi being 'stolen' into private domestic spaces, it also results in the public Wi-Fi space being accessed when the physical space was literally 'locked' or inaccessible at night time. Public space has traditionally been equated with physical access; so by locking the park gates at night it becomes closed. But the way public Wi-Fi is being filtered to create private bubbles of access beyond the spatial and temporal boundaries of the physical space.

Wi-Fi territories

Wi-Fi nodes or routers, which are essentially black box transmitters operating on the frequency of the 802.11b standard create a region of access of between thirty two metres indoors and ninety-five metres outdoors (depending on the technical characteristics of the router and also the physical features and materials of the immediate environment). However the Wi-Fi coverage over an area does not decrease evenly with respect to the source and Wi-Fi performance decreases roughly quadratically as the range increases at constant radiation levels. Due to the extent of this region, or territory, wireless access to the node can be available well beyond the physical borders and thresholds that traditionally delineate the boundary between private and public space. The setting in which the node operates exists on multiple levels; the technological setting of the node itself, the spatial setting in which it is accessible, and the social setting in which the user interacts with the technology.

Wi-Fi signal 'bleeding', where the signal extends beyond the physical boundaries of a building or area, complicates the relationship between ownership of a space and ownership of the Wi-Fi signal. Where public Wi-Fi is linked to the aspiration to support 'public-ness' then spatially it can be problematic when the boundaries of the public Wi-Fi do not correlate with the material boundaries of the public space. Similarly it can 'bleed' boundaries between work and home and between inside and outside. This occurs because the digital 'range' of the Wi-Fi signal is not confined by the spatial edges of rooms and buildings. In a study of Wi-Fi users in a public space within the Boundless network, cognitive mapping was used to establish how people understood the spatial boundaries of the Wi-Fi node and the corresponding boundaries of the café environment. The café consisted of a central seating and serving area, adjacent to a foyer and linked to an external courtyard, yet the public Wi-Fi range extended through the café space and out into the courtyard and also into the foyer, as well as spilling out into the street. Of fourteen people interviewed, twelve described availability of Wi-Fi as being confined within the physical territory of the indoor café space. Only two participants indicated the actual condition that the wireless access extended to cover an area not confined to the physical boundaries of the space. For instance one participant reported: 'well I know you can get access out in the foyer, but there's nowhere to sit out there so it's not usable' (Willis, 2008). This framing of behaviour also extended to other practical considerations, such as the suitability of tables for laptops, and in particular the location of electrical plugs. Participants equated the boundary of the Wi-Fi, not with the area it technologically covered, but with the behaviour it allowed. Despite being literally larger in extent than the physical 'edges' of the café, Wi-Fi users saw the boundary of the space based on the combined properties of the two types of spaces; the technical characteristics of the Wi-Fi range and the physical extents of the café walls.

Wi-Fi access and digital divides

At an urban scale, projects such as Taipei Free Wi-Fi (Taipei Government) Barcelona (Barcelona free Wifi) Tallin, Estonia (Wifi in Estonia), New York, USA (Citywide IT Services: NYCWiN), Wireless Antwerpen (Wireless België), Draadlos Groningen (Draadloos Groningen, 2015), Berlin (Visitberlin.de) and Toronto Canada have established municipal Wi-Fi infrastructures to provide broadband internet access in public places. Numerous other cities offer free Wi-Fi access in public or municipal buildings and increasingly in parks and central streets, squares and other public spaces. At a municipal level, libraries have been one of the key public spaces to offer free public Internet access (Bertot, McClure and Jaeger, 2008). According to a US study on public libraries and the Internet, over ninety-nine per cent of public libraries offer public internet access. At a smaller regional scale, the 2007 study found that public libraries are the only provider of free public Internet access in nearly seventy five per cent of US communities. In developing countries telecentres have proved a valuable equivalent to a library as a publicly accessible space for Internet access. Telecentres are 'a public place where people can access digital technologies and the Internet, information, and support and services that enable them to create, learn, play, and work – while building skills and connecting with others' (Wikipedia). They vary in size, facilities and services, ranging from a basic telecommunication service such as phone shops to fully interactive Internet-based training. Both libraries and telecentres address issues of access to Internet for those who may have either limited computer skills or lack of home-based Internet access. These groups represent those seen as being within the 'digital divide' due to inequality in their access to, use of, or knowledge of ICTs. According to the Tinder Foundation, seventy-five per cent of people counted as socially excluded in UK are also digitally excluded (UK Online Centres, 2008, p. 4). In the seventy-five per cent of US communities where the library is the only provider of free Wi-Fi access, and the numerous telecentres worldwide, the approach is that social and geographic boundaries can be countered through Wi-Fi access at such public spaces. For instance, a survey of telecentres in South Africa found that personal computers and the Internet were severely underutilised (Khumalo, 1998). This was found to be due to illiteracy and computer illiteracy, lack of awareness and culture about the use and benefits of ICT; the high cost of Internet connection through long-distance calls due to lack of local points-of-presence (POPs); and poor quality telecom connections (Latchem and Walker, 2001, p. 5). In a large scale study of telecentres, of the users who reported their income was less than the minimum wage, over seventy eight per cent used telecentres to access the internet, whereas in the high income brackets it was under thirty per cent (Gomez and Camacho, 2011, p. 20). As well as overcoming social boundaries, access to Internet can also counter geographical divides, particularly in isolated rural communities. For instance a case study by the Tinder Foundation of the provision of ICT access in regional libraries in rural settings, a seventy two year old participant reported how 'It can make such a difference to your life, especially if you live in an isolated area like us. I can drive at the moment, but I might not be able to in three or four years and we still need to do our shopping. Now I know how to use the internet I can do all that online' (UK Online Centres, 2008, p. 27).

Municipal Wi-Fi, and Internet access delivered through public telecentres and libraries is one route to delivering more equal and wider access. An alternative model is through community meshworks, which involves a peer-to-peer shared network of nodes linked up in a mesh framework. The Boundless network was an early pioneer of a community mesh, which was 'developed by local volunteers committed to developing a resource for the community. The primary focus is not on efficiency or economic development, but on providing a service that offers value to a local community, starting with free internet access' (James Stevens, personal communication 11 July 2014). The most successful meshwork is Guifi.net, with 26,885 nodes, which comprises of a user-owned, open and neutral network in which a growing community of volunteers can connect their computers to form a sort of intranet and, at the same time, share an Internet connection. The non-profit network is free, minus any individual costs for networking equipment, and anyone is allowed to join and use it how they want. Meshworks enable Wi-Fi access that parallels commercial and municipal Wi-Fi in terms of provision of access, but the underlying framework is different on a number of levels. Firstly, they create a distributed and decentralised network, rather than a spoke and hub set of connections, with no central control. But the key difference is that it requires sturdy social links to make the technical connections work. According to James Stevens, one of the founders of Boundless; 'we have a network of fifty nodes which link together and redistribute any broadband connectivity nodeholders have to offer. (Yet) it's the people who are most the important, and who require the most support, attention and encouragement. To establish a broad range of coverage in an area requires clear line of sight roof to roof connections but the ground level solutions are the hardest to co-ordinate and sustain' (James Stevens, personal communication 11 July 2014). A founder of a mesh network in Athens echoes these findings: 'It changes attitudes … people start sharing a lot. They start getting to know someone next door – they find the same interests; they find someone to go out and talk with' (Thompson, 2013). The linking up of a community through a public meshwork also results in the formation of social connections. By creating a shared public network and disrupting territorial models of individual ownership and private control, mesh Wi-Fi can establish new spatial and social boundaries within public space.

3.6   SUMMARY

Boundaries and thresholds mediate important spatial distinctions between inside and outside, public and private. These frameworks are ones that we use to create divisions and connections between one place and another and to differentiate the types of functions that can take place in a building or a space. These tend to rely on core notions based on spatial separation and proximity. Networked interactions also operate on frameworks that are no longer defined by near and far, in or out, and visible or invisible but by on/off modulations. One of the ways that buildings modulate our presence within them or exclusion from them is through sight; windows and doors define who enters or leaves and also create separation or connection between one space and another. For example, a large window will frame visual access to a beautiful view but exclude the wind or rain, and an exit will lead people out into the street, but can be locked to deny access. The way we connect or link in networked spaces is, no longer defined only by visual and physical access but by informational access. When access moves from physical access to informational access, the way in which people gain access to spaces also changes. When private space is no longer defined solely as control over a geographic domain; it is transformed into control over access and production of data within flexible information flows. The edges and thresholds of physical places are not only the material thresholds or bricks and glass, but also the digital territories of connections through mobile media and Wi-Fi. In the case study I looked at the example of Wi-Fi and discussed how it was part of the breaking down of boundaries in everyday lives; such as the distinction between work and home and also patterns of what constitute public and private space. This also has broader implications for how open access to network technologies in public space can counter digital divides, and the broader issue of the importance of places being both digitally and spatially 'open' or accessible.

REFERENCES

Ads of the World (2012) *Kit Kat: Free no WiFi zone.* http://adsoftheworld.com/media/ambient/kit_kat_free_nowifi_zone (Accessed: 21 August 2014).

Altman, I. (1975) *The Environment and Social Behaviour.* Monterey, CA: Brooks/Cole.

Barcelona free Wifi. http://www.bcn.cat/barcelonawifi/en/ (Accessed: 1 October 2014).

Benjamin, W. (ed.) (2002) *The Arcades Project.* New York: Belknap Press.

Berger, W. (1999) 'Lost in Space.' *Wired*, February 1999. http://www.wired.com/1999/02/chiat-3/ (Accessed: 23 October 2015).

Bertot, J., McClure, C. and Jaeger, P. (2008) 'The impacts of free public Internet access on public library patrons and communities.' *Library Quarterly*, 78, pp. 285–301.

Bourdieu, P. (1992) *The Logic of Practice.* Cambridge: Polity Press.

boyd, d. (2010) 'Social Network Sites as Networked Publics: Affordances, Dynamics, and Implications', in Papacharissi, Z. (ed.), *A Networked Self: Identity, Community, and Culture on Social Network Sites.* Oxford and New York: Routledge, pp. 39–58

boyd, d. (2011) 'Debating Privacy in a Networked World for the WSJ.' *Techplicy.com*. November 29. http://www.techpolicy.com/Blog/November-2011/Debating-Privacy-in-a-Networked-World-for-the-WSJ.aspx (Accessed: 5 August 2014).

boyd, d. (2014) *It's Complicated: The Social Lives of Networked Teens.* London and New Haven, CT: Yale University Press.

Bryant Park Corporation (2014) *Bryant Park Wireless Network.* http://www.bryantpark.org/plan-your-visit/wireless.html (Accessed: 1 October 2014).

Citywide IT Services: NYCWiN. http://www.nyc.gov/html/doitt/html/citywide/nycwin.shtml (Accessed: 1 October 2014).

Croft, J. (n.d.) *An FAQ on the LETS system.* http://www.gdrc.org/icm/lets-faq.html (Accessed: 1 October 2014).

Crow, B. and Miller, T. (2006) *Community Wireless Infrastructure Research Project1: Île Sans Fil Case Study Map.* http://www.cwirp.ca/files/CWIRP_ISF_case.pdf: Ryerson University.

Dodge, M. and Kitchen, R. (2004) 'Flying through code/space: the real virtuality of air travel.' *Environment and Planning A*, 36 (2), pp. 195–211.

Draadloos Groningen. (2015). http://draadloosgroningen.nl/wordpress/ (Accessed: 1 October 2014).

Drucker, J. (2013) 'Google Joins Apple Avoiding Taxes With Stateless Income.' *Bloomberg*. http://www.bloomberg.com/news/2013-05-22/google-joins-apple-avoiding-taxes-with-stateless-income.html, 22 May 2013.

Forlano, L. (2009) 'WiFi Geographies: When Code Meets Place'. *Inf. Soc.*, 25 (5), pp. 344–352.

Fujimoto, K. (2005) 'The Third-Stage Paradigm: Territory Machines from the Girls' Pager Revolution to Mobile Aesthetics', in, Ito, M., Okabe D. and Matsuda, M. (eds), *Personal, Portable, Pedestrian: Mobile Phones in Japanese Life*. Cambridge, MA: The MIT Press.

Furuto, A. (2011) '"Urban Future" at Design Miami 2011 / BIG + Kollision + Schmidhuber & Partner'. *Arch Daily* (21 August 2014).

Gibson, J. (1979) *The Ecological Approach to Visual Perception*. Boston, MA: Houghton Mifflin.

Goffman, E. (1963) *Behaviour in Public Places: Notes on the Social Organization of Gatherings*. New York: Free Press.

Goffman, E. (1990) *The Presentation of Self in Everyday Life*. London: Penguin.

Gomez, R. and Camacho, K. (2011) 'Who Uses Public Access Venues?', in, Gomez, R. and Camacho, K. (eds), *Libraries, Telecentres, Cybercafes and Public Access to ICT: International Comparisons*. Hershey, PA: IGI Global, pp. 11–22.

Gordon, E. and de Souza e Silva, A. (2011) *Net Locality: Why Location Matters in a Networked World*. Chichester, UK: Wiley-Blackwell.

Gottdiener, M. (2001) *Life in the Air: Surviving the New Culture of Air Travel*. Lanham, MA: Rowman and Littlefield.

Graham, S. and Marvin, S. (1996) *Telecommunications and the City*. London: Routledge.

Habuchi, I. (2005) 'Accelerating Reflexivity', in, Ito, M., Okabe, D. and Matsuda, M. (eds), *Personal, Portable, Pedestrian: Mobile Phones in Japanese Life*. Cambridge, MA: The MIT Press.

Hall, E. (1966) *The Hidden Dimension*. Garden City, NY: Doubleday Anchor Books.

Hampton, K., Livio, O. and Sessions, L. (2010) 'The Social Life of Wireless Urban Spaces: Internet Use, Social Networks, and the Public Realm'. *Journal of Communication*, 60 (4), pp. 701–722.

Hannam, K., Sheller, M. and Urry, J. (2006) 'Editorial: mobilities, immobilities and moorings'. *Mobilities*, 1 (1), pp. 1–22.

Heathrow Airport *Phones – public, rental and charging*. http://www.heathrowairport.com/heathrow-airport-guide/services-and-facilities/phones-and-charging (Accessed: 1 October 2014).

Hillier, B. and Hanson, J. (1984) *The Social Logic of Space*. Cambridge: Cambridge University Press.

Höflich, J.R. (2006) 'The mobile phone and the dynamic between private and public communication: Results of an international exploratory study'. *Knowledge, Technology & Policy*, 19 (2), pp. 58–68.

Humphreys, L. and Liao, T. (2013) 'Foursquare and the Parochialization of Public Space'. *First Monday*, 18 (11).

Impact Hubs (2014) *Impact Hubs*. http://www.impacthub.net/what-is-impact-hub (Accessed: 21 August 2014).

Ito, M. (2012) 'Introduction', in Varnelis, K. (ed.), *Networked Publics*. Cambridge, MA: The MIT Press.

Ito, M. and Okabe, D. (2005) 'Technosocial situations: emergent structuring of mobile e-mail use', in Ito, M., Okabe, D. and Matsuda, M. (eds), *Personal, Portable, Pedestrian: Mobile Phones in Japanese Life*. Cambridge, MA: The MIT Press.

Jacobs, J. (2002) *The Death and Life of Great American Cities*. New York: Random House.

Katz, J.E. and Aakhus, M. (eds) (2002) *Perpetual Contact: Mobile Communication, Private Talk, Public Performance*. Cambridge: Cambridge University Press.

Kelleher, K. (2013) 'Mobile growth is about to be staggering'. *Fortune*. http://fortune.com/2013/02/20/mobile-growth-is-about-to-be-staggering/ (Accessed: 23 October 2015).

Kitchin, R. and Dodge, M. (2011) *Code/Space: Software and Everyday Life*. Cambridge, MA: The MIT Press.

Lasen, A. (2003) 'A comparative study of mobile phone use in public places in London, Madrid and Paris'.

Latchem, C. and Walker, D. (2001) *Telecentres: Case Studies and Key Issues*. Vancouver: The Commonwealth of Learning.

Latour, B. (2005) *Reassembling the Social: An Introduction to Actor-Network-Theory*. Oxford: Oxford University Press.

Law, J. (1994) *Organizing Modernity: Social Order and Social Theory*. Oxford: Blackwell.

Ling, R. (1997) *'One Can Talk about Common Manners!': The Use of Mobile Telephones in Inappropriate Situations*. Stokholm: Telia.

Lynch, K. (1967) *The Image of the City*. Cambridge, MA: The MIT Press.

Mackenzie, A. (2008) 'FCJ-085 Wirelessness as Experience of Transition'. *FibreCulture* (13).

Mackenzie, A. (2010) *Wirelessness*. Cambridge, MA: The MIT Press.

Meyrowitz, J. (1985) *No Sense of Place: The Impact of Electronic Media on Social Behavior*. New York: Oxford University Press.

Mitchell, W.J. (1995) *City of Bits*. Cambridge, MA: The MIT Press.

Mitchell, W.J. (2004) *Me++: The Cyborg Self and the Networked City*. Cambridge, MA: The MIT Press.

Morgenland. http://www.morgenland-berlin.de/ (Accessed: 1 October 2014).

n.n. (1922) 'Radio; Farm Service Growing'. *The New York Times*. http://query.nytimes.com/gst/abstract.html?res=990CE0D91239EF3ABC4850DFB1668389639EDE. (6), 30 July 1922.

Pertierra, R. (2005) 'If You Can't Afford a Room of Your Own, Buy a Mobile Phone'. *Conference on Mobile Communication and Asian Modernities*. Hong Kong, SAR: 8 June 2005.

de Sola Pool, I., Decker, C., Lizard, S., Israel, K., Rubin, P. and Weinstein, B. (1977) 'Foresight and Hindsight: The Case of the Telephone', in de Sola Pool, I. (ed.), *The Social Impact of the Telephone*. Cambridge, MA: The MIT Press, pp. 127–157.

Priest, J. (2004) *The State of Wireless London*. http://informal.org.uk/people/julian/publications/the_state_of_wireless_london/ (Accessed: 1 October 2014).

Rapoport, A. (1994) 'Spatial organization and the built environment', in Ingold, T. (ed.), *Companion Encyclopedia of Anthropology: Humanity, Culture and Social Life*. Abingdon: Routledge, pp. 460–502.

Rayner, R. (1998) 'Nowhere, U.S.A.', *The New York Times*. http://www.nytimes. com/1998/03/08/magazine/nowhere-usa.html, 8 March 1998.

Riley, T. (1999) 'The Un-Private House'. *The Un-Private House*. New York: The Museum of Modern Arts, New York.

Schmidt, M. and Schmitt, E. (2013) 'Obama's Portable Zone of Secrecy (Some Assembly Required)'. *The New York Times*, 9 November 2013.

Shepard, M. (ed.) (2011) *Sentient City: Ubiquitous Computing, Architecture, and the Future of Urban Space*. Cambridge, MA: The MIT Press.

Simmel, G. (1994 [1909]) 'Bridge and Door'. *Theory, Culture & Society*, 11, pp. 5–10.

Stevens, Q. (2007) 'Betwixt and Between: Building Thresholds, Liminality and Public Space', in Franck, K. and Stevens, Q. (eds), *Loose Space: Possibility and Diversity in Urban Life*. Abingdon and New York: Taylor & Francis, pp. 73–92.

Taipei Government *Taipei Free Wifi*. http://www.tpe-free.taipei.gov.tw/tpe/index_en.aspx (Accessed: 1 October 2014).

The Economist (2013). *The rise of the sharing economy*. http://www.economist.com/news/ leaders/21573104-internet-everything-hire-rise-sharing-economy (Accessed: 1 August 2014).

Thompson, C. (2013) 'How to Keep the NSA Out of Your Computer: Sick of government spying, corporate monitoring, and overpriced ISPs? There's a cure for that'. *Mother Jones*. http://www.motherjones.com/politics/2013/08/mesh-internet-privacy-nsa-isp (Accessed: 1 October 2014).

Thrift, N. and French, S. (2002) 'The Automatic Production of Space'. *Transactions of the Institute of British Geographers*, 27 (3), pp. 309–335.

UK Online Centres (2008) *Digital inclusion, social impact: a research study*. http://www. tinderfoundation.org/sites/default/files/research-publications/digital_inclusion_ research_report.pdf: Tinder Foundation.

Varnelis, K. (ed.) (2012) *Networked Publics*. Cambridge, MA: The MIT Press.

Visitberlin.de *W-LAN for all: "Public Wi-Fi Berlin": New Hotspots in the Capital*. Berlin Tourismus & Kongress GmbH (Accessed: 1 October 2014).

van de Vliet, V. (2013) 'Space for People, Not for Cars'. *Works That Works 1*. https://worksthatwork.com/1/shared-space (Accessed: 1 October 2014).

Warschauer, M. (2003) *Technology and Social Inclusion: Rethinking the Digital Divide*. Cambridge, MA: The MIT Press.

Wellman, B. (2001) 'Little Boxes, Glocalization, and Networked Individualism', in Tanabe, M., van den Besselaar, P. and Ishida, T. (eds), *Second Kyoto Workshop on Digital Cities II, Computational and Sociological Approaches*. London: Springer-Verlag, pp. 10–25.

Westin, A.F. (1967) *Privacy and Freedom*. New York: Atheneum.

Wifi in Estonia. (2014). http://www.visitestonia.com/en/things-to-know-about-estonia/facts-about-estonia/wifi-in-estonia (Accessed: 1 October 2014).

Wikipedia (n.n.) *Criticism of Facebook*. Wikipedia, http://en.wikipedia.org/wiki/Criticism_of_ Facebook (Accessed: 1 October 2014).

Wikipedia *Telecentre*. Wikipedia. http://en.wikipedia.org/wiki/Telecentre (Accessed: 1 October 2014).

Willis, K.S. (2006) *Wayfinding Situations*. Bauhaus University Weimar.

Willis, K.S. (2012) 'Being in Two Places at Once', in Abend, P., Haupts, T. and Müller, C. (eds), *Medialität der Nähe: Situationen – Praktiken – Diskurse*. Bielefeld: Transcript.

Winston, B. (1998) *Media Technology and Society: A History From the Telegraph to the Internet*. London: Routledge.

Wireless België. (2014). http://www.wirelessantwerpen.be/ (Accessed: 1 October 2014).

Zickuhr, K. and Smith, A. (2012) *Digital differences*. http://www.pewinternet.org/2012/04/13/digital-differences/ – fn 188–20: Pew Research Internet Project.

4

Publics

4.1 NETWORKED PUBLICS

Playing the city

4.1 Digital photo of 'sighting' sent as part of 'Can You See Me Now?' mixed reality game.

I have a digital photo of a 'sighting' of my father that is saved somewhere within the files on my computer (Figure 4.1). Except it's not him in the photo; instead there is a picture of a pavement which shows a bit of building façade in a street with people walking past. It's a photo of a place I don't recognise, as I have never actually been there. I was never there, and my father actually passed away about a year before the photo was taken. The photo was taken by a runner in the Blast Theory game in Sheffield called 'Can You See Me Now?' (Blast Theory). When I registered for the game through an online platform as a virtual player I was prompted to answer the question: 'Is there someone you haven't seen for a long time that you still think of?' From then on I was strangely aware of the unlikely cast of players in the game; myself, sitting at my laptop in a room in London, a virtual me on the screen, other virtual players, the real runner (an actor) out on the streets of Sheffield, and the ghostly presence of my father propelled to be the unlikely 'protagonist' in the game. The game play took place live in the city centre of Sheffield, with real runners equipped with GPS-enabled phones whose presence was tracked on an online platform that showed their location on a map of the city. On the screen I used the arrow keys to move the virtual me out of the path of the runner, as they chased me down. There was only one runner and about twenty players. When the runner became distracted by other players in the game I hid in the virtual grid of the screen-based map of the space. An audio stream from the runner's walkie-talkies meant I could hear, but not see, the runner: the audio feed played the sound of their breathing as they ran against a background of the sounds of the city. There was also friendly banter exchanged back and forth between the other players in the game; insults, teasing, goading and humour creating a shared collective sense of space between the remotely located players. We were all part audience, part player. Meanwhile, the physical effort of the actors or runners was palpable; out of breath, their spoken words revealed the tension and challenge of the game in the streets and buildings of real space. After about twenty minutes of virtual running, the runner caught me. Or, rather, the virtual me coincided with the location of the runner on the virtual map. My father's name was briefly spoken aloud by a runner on the distant streets of the city and existed for a second before fading into the ether. I then received a 'sighting'; an uploaded photo of the location of the runner at the moment when I was caught; the unfamiliar street full of strangers. The photo is strangely dehumanised since it is taken from an odd height (about the level of the view from a ten year-old child), probably due to the fact that the photo was automated and the runner did not raise the mobile device up to eye level to take the image. The strange viewpoint of the photo typifies much of the experience of 'Can You See Me Now?' for there was both a sense of online and offline presence but also an awareness that what you saw was a performance in which you were in a space somewhere between being actor and audience.

In fact, Blast Theory, the group behind the urban game describe their approach as 'performance that mixes audiences across the internet, live performance and digital broadcasting' (Blast Theory). Blast Theory are part of a growing number of artistic and commercial 'directors' who choreograph participatory events in the city that set up new configurations of actor and audience; where the streets and spaces of the city become a stage for an unravelling theatre. The public space of the street has always been a site of public performance, of scene's unfolding with a cast of actors and an audience. Now the buildings and the media of mobile phones and media facades are creating a new mix of performative public space.

Streets, squares and cafés

According to Whyte, the street is 'the river of life of the city, the place where we come together, the pathway to the center' (1988, p. 9). The street is social theatre, and it needs people to fill it in order to contribute to the life of the city. Jacobs also highlighted the value of street life for a city and outlined how 'the sidewalk must have users on it fairly continuously … . Nobody enjoys sitting on a stoop or looking out a window at an empty street. Almost nobody does such a thing. Large numbers of people entertain themselves, off and on, by watching street activity' (2002, p. 35). Being together with others in the city is not about a domestic sense of togetherness, where people are known to each other. Simmel characterises the street as made up of strangers, and that these strangers are both 'close to us, insofar as we feel between him and ourselves common features … (and) far from us, insofar as these common features extend beyond him or us, and connect us only because they connect a great many people' (1971, p. 143). Simmel argues that this aspect of metropolitan life is the cause of strangeness, since the 'feelings of isolation are rarely as decisive and intense when one actually finds oneself alone as they are when one is a stranger among many physically close persons, at a party, on a train, or in a city' with the consequence that 'one nowhere feels as lonely and lost as in the metropolitan crowd' (1950, p. 117). But the exposure to strangers is how Lofland defines value in the 'public realm'. It is areas of the city in 'which individuals in co-presence tend to be personally unknown or only categorically known to each other' (Lofland, 1998, p. 9), and Sennett maintains that 'the public realm can be simply defined as a place where strangers meet' (2010, p. 261). A key factor that enables people to encounter others in public is that the space is equally open to all. It is this accessibility that means that the public realm is a 'natural stage' and a powerful medium of communication (Lofland, 1998, p. 124). These approaches see the public space as a place of gathering and social encounter rather than as spaces for moving through. They support a performance of sociality among strangers that help to constitute an idea of public-ness. According to Ballantyne this is a key characteristic of architecture since 'most scholars employ the first notion of architecture as performance, in which a building and its interior are seen to function as a public theatre' (2002, p. 25). The 'architecture as performance' approach privileges a way of thinking about architecture that sees buildings and public space, not as fixed and complete, but as scenery or a platform that is only brought to life through the theatre of its actors and audience; that is people.

Digital and urban publics

The fate of the public realm or the urban commons has been a subject of broader debate in urban theory over a number of decades (Gehl, 1987; Harvey, 2012; Jacobs, 2002; Sennett, 1977; Whyte, 1980). A key point of discussion is the apparent demise of public space and a corresponding retreat into private or privatised spaces. According to Harvey this is a result of the recent effects of 'privatisations, enclosures, spatial controls, policing, and surveillance upon the qualities of urban life in general' (Harvey, 2012, p. 67). Harvey points out that the public realm has always been an unstable and malleable social relation, but argues that the challenge is how society can exploit this potentiality 'to build or inhibit new forms of social relations (a new commons) within an urban process' (2012, p. 67). One of the factors that is seen as contributing, or at least reflecting, the loss of the public realm is the proliferation of media in public spaces. Gehl puts the fault directly at the feet of new media and argues that 'the telephone, television, video, home computers, and so forth have introduced new ways of interacting. Direct meetings in public spaces can now be replaced by indirect telecommunication. Active presence, participation, and experience can now be substituted with passive picture watching, seeing what others have experienced elsewhere … Something is missing' (Gehl, 1987, p. 49). The privatisation of public space is seen by many as directly linked to the devices and screens that create personal, privatised spaces in public ones. The argument is that the stages necessary for public life to take place are being colonised by multiple and isolated private encounters enabled by fixed and mobile media (Höflich, 2005, p. 124). This happens on a number of levels. At the scale of the built world, there is the introduction and spread of 'urban screens' (Struppek, 2006) or 'moving billboards' (Manovich, 2006), many of which deliver advertising and commercial information. According to McQuire 'the migration of electronic screens into the external cityscape has become one of the most visible tendencies of contemporary urbanism' (2006). At the scale of the portable screen, Sheller and Urry state that more recent developments in the use of mobile media mean that 'the proliferation of screens, from the miniature ones displaying text messages on handheld devices to the large ones in public spaces is allowing for new kinds of informational mobilities that use public spaces for "private" purposes' (2000). If architecture creates a stage in the public realm that allows people to access a common space for gathering, meeting and interaction, then it must follow that media should support the creation of a commons that is participatory, whether actively or passively. The challenge is to find ways to extend the hybrid mix of media and architecture to something that is more than the sum of both. Many architects and interior designers have actively embraced electronic media, but 'they typically think of it in a limited way; as a screen i.e. as something that is attached to the 'real' stuff of architecture – surfaces defining volumes' (Manovich, 2006, p. 236). The challenge is whether urban and augmented screens can become tools to contribute to urban publics that actively involve their audience.

(Inter)active architecture?

What roles do media and architecture have in promoting other forms of spatial agency that contribute to the creation of urban publics or a commons? In contrast to a passive screen culture of television, the interactive capacity of the screen, either in the hands of a citizen or where their content is accessible to the public suggests new modes of participating. It also suggests new models of audience and performance. When embedded or linked to the architecture of public space it shifts the idea of the passive opening and surface from that where 'a simple, well-placed opening frames a view, regulates the flow of sound, light and air and dramatises passage from one space to another' to a screen that 'means something different, and often more active than the frame' (McCullough, 2014, p. 154). Their digital nature makes these 'screening platforms' an experimental zone on the threshold of virtual and urban public space (Struppek, 2006, p. 174).

This shift to a screen-based culture is materialised at a range of scales. Firstly, the transfer of the fixed computer to the mobile and portable screen of the phone, laptop and tablet meant public space became infused with devices. Ito argues that mobile media does not necessarily equate with the privatisation of public space and offers up 'networked publics' (boyd, 2010) as an alternative to terms such as audience or consumer (Ito, 2012). According to Ito, this approach moves beyond a simple on or off engagement of either passive or consumptive, and foregrounds a more subtle approach to engagement. This is necessary because 'now publics are communicating more and more through complex networks that are bottom-up, top-down, as well as side-to-side. Publics can be reactors, (re)makers and (re)distributors, engaging in shared culture and knowledge through discourse and social exchange as well as through acts of media reception' (Ito, 2012, p. 3). The second way that screen based culture affects public space is where the facades and surfaces of the urban environment are increasingly overlaid or replaced with either projected or embedded screens. Struppek termed these 'urban screens' and defined them as 'digital moving displays with a new focus on supporting the idea of urban space as a space for the creation and exchange of culture and the formation of a public sphere using criticism and reflection' (Struppek, 2006). According to McQuire urban screens open up new models of participation and interaction in public space since 'unlike cell phones or MP3 players which tend towards individual forms of consumption – 'mobile privatisation … – they are oriented towards collective forms of engagement' (2006). This requires a reflection on how architecture, networks and their devices can work together as a process that sees the creation of publics as a performance with a shifting cast of public and private actors.

Aims

The topic of what constitutes the public realm and how it can be supported and promoted has been a source of much discussion and debate in architecture and urban theory. The introduction into public space of a range of networked technologies and the practices associated with them further raises challenges for the construction of publics. This chapter explores ideas of agency and performance in the built environment, with a focus on urban public space. Working from the historical context of television and broadcast media, concepts of actor and audience, as well as frameworks for how people can participate in the public realm when it is experienced through digital technologies are examined. In the case study I focus on urban screens, which is a term that refers to screen-based interactions and experiences either fixed or mobile. Through these examples, different conditions for the construction of publics are explored; whether through exposure to strangers through performative urban screen-based encounters, through shared encounters with others or through playful interventions into existing societal rules created by mixed reality games.

4.2   PERFORMING SPACES

Networked publics and the public realm

If we take Lofland and Sennett's reading of the public realm; as essentially an open platform where strangers can meet and experience co-presence, then the public realm is principally a social entity, and is about opening up channels of communication. Arendt states that one of the ways we can experience the public realm is through exposure to 'the presence of others who see what we see and hear what we hear' as this 'assures us of the reality of the world and ourselves' (1999, p. 50). She also makes the observation that the public realm is distinct from the private realm; in that 'the term "public" signifies the world itself, in so far as it is common to all of us and distinguished from our privately owned place in it' (Arendt, 1999, pp. 51–52). Jacob's recognition of the importance of 'the eyes on the street' (2002, p. 45) and Whyte's highlighting of the need for people to attract other people in urban plazas (1988) represent the hard evidence of the social value of people gathering and being aware of one another in urban settings. This helps to contribute to what Putnam terms 'social capital' (2000) and is directly linked to the amount of trust and 'reciprocity' in a community or between individuals. In his book *Bowling Alone*, Putnam exonerates new media and the internet of responsibility for facilitating the demise of the public realm, which he links more clearly to the TV viewing culture and the rise of the two working parent family (2000). Instead Putnam highlights the potential of internet-based networks for social connectedness and civic engagement (2000, p. 174). If network technologies are about social connection then the argument for their contribution to the demise of the public realm can be linked to the way that they affect face-to-face interaction.

According to Goffman, meeting face-to-face enables people to develop encounters, display attentiveness and commitment (1963, p. 92). This means that public space must be an inclusive and accessible space for communication and interaction for all present. When looking at how networked technologies fit into this arena, one of the key factors is the nature of presence and co-presence, since if the public realm requires people to be present with one another then mobile media and the Internet raise significant questions about their ability to contribute to publics. However a number of studies have found that the synchronicity or intensity of mediated communication constructs an awareness of others (Arminen and Weilenmann, 2009; Licoppe, 2004). The binary of either private or public is also dissolved by practices of mobility and media use. Instead of either being in public or in private many people move between these conditions either physically or through the use of media much more fluidly. As Mitchell highlights 'there is a strong relationship between prevailing network structure and the distribution of activities over public and private places ... where networks go wireless, they mobilise activities that have been tied to fixed locations and open up ways of reactivating urban public space' (2004, p. 158). For instance Ling and Yttri (2002) documented the use of mobile phones to support meetings where collaboration is ad-hoc and the core interaction is not necessarily face-to-face. These spaces of collaboration show a fine balance between remote interaction and face-to-face meeting, a practice that has also been referred to as 'zooming with the feet' (Bertelsen et al., 2001). This practice of acting remotely but then coming together for a specific purpose is also highlighted by Ito (2005), in the context of mobile media use, who terms it a 'flesh meet'. The use of mobile media in public space certainly complicates the condition of being 'in public', but it certainly does not mean that publics are disappearing. It is more a question of whether public space is the only stage where 'publicness' can be enacted or whether networked spaces also create the conditions for people to experience the public realm.

Actor, audience and stage

The public realm is about creating the conditions for people to participate in 'being social'. In boyd's concept of networked publics, 'the ways in which technology structures them introduces distinct affordances that shape how people engage with these environments. The properties of bits – as distinct from atoms – introduce new possibilities for interaction. As a result, new dynamics emerge that shape participation' (boyd, 2010, p. 39). The mixing of technology and public space is already happening. Smart and mobile phones, animated advertising screens, transport information boards and social media augment the experience of public space in the city. This raises questions of how the framework for such a social space is produced. Struppek sketches out two scenarios for how citizens might shape public space and media; one negative and one positive. The first (negative) scenario is the top-down, driven by city planners and based on an '"event culture", the "creative city" shaped by the creative class or the "festivalisation" of public space with festivals entertaining all year long'. In this scenario the city becomes a theatre stage with carefully produced and enjoyable infotainment in line with what Debord foresaw in 1967 in the 'Society of the Spectacle' (Struppek, 2014, p. 2). The second (positive) scenario is a bottom-up counter strategy of 'interactive installations, participatory sound-recordings and communication-sculptures, public (guerrilla) projections or mobile screen interventions, social, playful light and sound-triggering installations, public community message boards and space annotations discovering local cultures, Wi-Fi art and psycho-geographic performances as well as location-based mobile gaming and outdoor mixed-reality games' that contribute to 'the revival of citizenship' (Struppek, 2014, pp. 3–4). The key factor that differentiates the projects, in terms of the two models of the public realm they envisage, is linked to the question of who is the actor and who is the audience. In the first scenario the city planners are the actors and the citizen the audience, whereas in the second, the citizen is both actor and audience. McCullough (2014) suggests a less active role of an actor in the practice of visual 'foraging', or a way for the citizen to actively make choices about where to focus their attention. Again, as a counter strategy to visual consumption he advocates that 'with so many new relations among windows, screens frames and facades now filling the everyday space, watching has become less important, and foraging has become more so … context and sensibility intertwine' (McCullough, 2014, p. 164). Whether simply foraging or actively participating in the performance in public space, these approaches require an architecture that moves beyond the 'surface as electronic screen paradigm' (Manovich, 2006). According to Manovich 'architects now have the opportunity to think of the material architecture that most usually preoccupies them and the new immaterial architecture of information flows within the physical structure as a whole … . In other words, architects along with artists can take the next logical step to consider the 'invisible' space of electronic data flows as substance rather than just as void' (Manovich, 2006, p. 237). This requires architecture to engage with its role, not as a static producer of urban structure, but as a stage where the public realm can be enacted, and where participatory media is part of the performance.

Agency and participation

Ever since the beginnings of the internet and the emergence of a distinction between identity online and offline, one of the issues with networked media has been how, when and where people reveal their 'real' identity. For Arendt one of the criteria for the public realm is that 'a public space must enable individuals to be visible before others, to reveal and perform their "public selves", in order to be a space for the realisation of their identity and uniqueness' (Thuma, 2011). Arendt understands public space as a kind of 'stage' for the short-lived performances of the individual. That is, they are public spaces that give some kind of reality and durability to human life and to the world that is created between individuals through their actions. As we become even further bound up with our devices and connections, the point at which we are mindful actors in enactment of behaviours and the degree to which we are merely reactants or even puppets start to blur. Urban media interfaces create a delicate balance between personal participation and collective interaction, between active engagement and reflective contemplation. But the issues underlying the role of agency and urban media have existed since the beginnings of the shared use of Internet based communication. These issues initially played out in the online environments that characterised early, desktop PC-based internet use, but as the use of remote media becomes increasingly mobile and ubiquitous the same issues of agency start to impact on public space; the new arena for disembodied co-presence. In urban public space the degree to which media affects agency is due to the fact that the remote nature of the communication blurs who is in control. A mobile phone ringtone or text notification demands a response regardless of the appropriateness of the location or availability of the recipient, creating a state which Licoppe refers to as the 'crisis of the summons' (2010). The communicative availability of a person to the 'summons' of a remote other is no longer under the control of the actor, but is in a sense manipulated by the remote other. Although Licoppe notes that 'users are becoming more skilled and turning into "pragmatic amateurs," less inclined to accept the imposition of a summons' (2010, p. 288) it is still important to consider issues of agency in relation to publics and public space. Moving beyond the scale of the individual to the scale of the collective is the actions of 'smartmobs', a term introduced by Rheingold (2003), to describe a group of people who use network connections with mobile communications, PCs, and the Internet to organise collective action. Smart mobs work most effectively in open and accessible public spaces with large numbers of people; just those that would be defined as best representing the public realm.

The question then is how can the static nature of our built environment and space start to respond to these forms of serendipity and micro-coordination of social behaviour? The term 'performative architecture' was introduced by Andrews and Taylor (1982) to describe architecture as a backdrop for body movement. Some uses of performative architecture are described by Kolarevic and Malkawi (2005) as either a building as a shell for activities, the reality of the building itself (how it is realised) or the relationship between these two aspects (i.e. the effect it has on the social and the ability to affect and transform the building itself). If architecture can contribute to the public realm it needs to perform in third meaning; that is it needs to both act on the social setting it is within and this needs to have some reciprocal expression in the way the architecture literally performs as a built space or structure. Concepts of interactivity and participation in architecture tend to be framed around either layers of media on the surface of individual buildings which can respond to some effect (this is a massive over simplification – justify, and reference). In this sense 'interactivity' is not simply a choice among a menu of predictable consequences, but belongs to a 'more open horizon in which contingency and unpredictability play a greater role' (McQuire, 2006). Latour and Yaneva characterise such an architecture as one in motion, and ask us to 'picture a building as a moving modulator regulating different intensities of engagement, redirecting users' attention, mixing and putting people together, concentrating flows of actors and distributing them so as to compose a productive force in time-space' (2008, p. 87). Architecture that performs in public space is never static; it can open up to series of possible outcomes channelled by the agency of multiple unknown actors.

4.3   HISTORICAL CONTEXT

Technological public spaces: from street lights to urban screens

Towards the end of the nineteenth century the streets of Europe and America underwent a transformation; electric lighting was installed. The existing gas lamps that had illuminated the streets, which had to be lit each day and tended-to individually, were slowly replaced with a system of electric streetlights. The introduction of electricity into the urban street had a range of implications for public life. The first was that it became possible for people to be out on the street at night. Prior to this the street was generally viewed as a dangerous space after nightfall, and to be out on the streets suggested involvement in some kind of impropriety (Schivelsbusch, 1995, p. 82). The second implication was that electricity brought with it an automation of the street lighting system. Gas lamps that had needed to be lit by a watchman every evening, were replaced by streetlights, which meant that the streets became part of a more homogenous, systematic and serviced 'public' space (Nye, 1990). The street also became simultaneously more accessible as a social space; by extending the time in which public activities could take place into the evening and night, but also less social in that it was no longer maintained and managed by people such as the light watchmen.

Whilst electric lighting was bringing private life out onto the stage of public space in the midst of the twentieth century, the introduction of the television in post-war society had the opposite effect. Public life started to be broadcast in the private sphere of the home, and private life was profoundly re-socialised by radio and televisions (Scannell, 1996, p. 141). The television changed the nature of the domestic space of the home, and connected it to other places and events, so that 'consequently the home is a less bounded ... environment because of family members' access and accessibility to other places ... through radio, television and the telephone' (Meyrowitz, 1985, p. vii). Initially, when televisions were still rare devices in the home, TV viewing was a semi-public event with friends and neighbours invited to join the viewing. But over the latter part of the twentieth century as part of what Moore refers to as the 'disembedding mechanism' (2000, p. 6), television viewing became an increasingly isolated and private experience. Williams (1974) referred to this as 'mobile privatisation'; a process through which urban dwellers, were increasingly physically isolated from each other in the solitude of their suburban living rooms, enjoying a growing virtual mobility due to the arrival of the television. However the private viewing of public events via the television created what Scannell defines as a 'doubling of place' (1996), where public events occurred both in the public place of the event and also where it was watched or heard. In bringing public events and spaces into the domestic sphere, television also brought a sense of 'liveness' or immediacy of public events, where a 'television broadcast is characterised as a performance in the present' (Auslander, 1999, p. 15). Indeed over the latter part of the twentieth century television as a mode of experiencing events increasingly moved out of the domestic sphere; into bars, restaurants, airports, sports venues and retail spaces. By the end of the twentieth century the liveness of television and the electrification of street lighting were starting to mutually contribute to a transformation in urban space. The television screen was changing from a framed screen located in a private domain to a viewing surface, as noted by McQuire who observed that 'the television set has morphed from a small-scale appliance – a material object, a piece of furniture primarily associated with domestic space – to become an architectural surface resident not in the home but in the street outside' (2006). Made possible by the technical development and availability of LED lighting and high-powered projection systems, large-scale urban screens became viable features within the built fabric. The emergence of large scale screens in public space is generally characterised by their use for broadcasting advertising; that is, viewing based on consumption, mirroring the commercialisation of television viewing. In the last decade we have seen a general increase in the introduction of fixed screens within the urban fabric but also the increasing ubiquitous presence of mobile screens on smartphones and tablets.

Social public spaces: from the agora to networked publics

It is only in the last century that the concept of public space emerged as a key component of civic life in the city (Habermas, 1989). Yet, the concept of the public sphere as a way of understanding how a society might hold the potential for collective thought or action has existed since Greek times. The agora was one of the first physical manifestations of the public sphere. The agora was the centre of civic activity, a place for citizens to gather and discuss 'in public' and also a square surrounded by public buildings. Arendt (1999) highlighted how the public sphere needed to be spatialised and the importance of public spaces to the human condition. Habermas (1999), outlined the public sphere as it emerged in the eighteenth century, and described how it was realised through a public space such as a coffee or tea house where people gather, read newspapers, talk and discuss common interests. Whether coffee house or agora, in the twentieth century the streets, public and semi-public spaces of the city became understood as a common stage for the social theatre of meetings and encounters. Yet, a whole range of theorists (Harvey, 2006; Jacobs, 2002; Low and Smith, 2006; Sennett, 1977) have argued that the public sphere that had characterised early modernism was displaced by a pervasive withdrawal into a private sphere in the post war era. In the late eighties and nineties a new challenge to public space was raised by theorists who argued that the use of mobile media and the Internet contributed to a privatisation of public space (Ling, 2005). The wider implication of the impact of these technological developments was that it 'made the withdrawal from participation in the public realm a genuine option' (Lofland, 1998, p. 144). The key argument was that the use of technologies such as mobile phones and social networking meant that social life was no longer played out in public space, since it took place without face-to-face interaction, and thus moved away from origins of the public sphere 'in the context of the café, the learned society, and the salon' (Ling, 2005, p. 16). Counter arguments, such as the discussion of 'networked publics' (boyd, 2010; Varnelis, 2012), suggest that publics are very much part of how people interact with the internet and mobile phones, but the public sphere may not only be enacted in public space. Instead, they argue, it happens across a whole range of platforms; spatial and technological. Hampton, Livio and Sessions, in a study of Wi-Fi use in public space make a similar point in that 'exposure to diversity of opinions and issues within the public sphere is dependent on the range of external inputs available from the mass media and everyday interactions embedded within the private, parochial, and public realms' (2010, p. 702). The growth in the public realm is instead being realised in parochial spaces that are characterised by 'a sense of commonality among acquaintances and neighbors who are involved in interpersonal networks that are located within communities' (Lofland, 1998, p. 10). This is seen as evidence that media use is changing the nature of public space, but still allowing the conditions for participation in the public sphere. McQuire, in his study of the effect of urban screens on public space also concludes that 'new forms of public interaction which involve sharing and negotiation between individual and collective agency can play a vital role in challenging the dominance of public space by spectacular 'brandscapes' or its pacification by surveillance' (McQuire, 2006). The implications are that if the new participatory models of participatory media are made accessible within public space then they can contribute to the public sphere by encouraging collective action and encounters between strangers.

4.4   NEW DIGITAL PUBLICS

There are a range of readings as to how the public realm can be supported through public space; through exposure to strangers (Arendt, 1999), through interaction and discourse with others (Habermas, 1999) and through exposure to diversity (Sennett, 1977), then the space in which these occur is important in terms of how it provides a stage for such publics to be enacted. The role of technology can be seen at once as reframing face-to-face interaction, but simultaneously enabling a much more diverse exposure to others through social networks and online communities. In the following section I review a series of ways in which technology may mediate the construction of publics; through new models of shared or co-production of space, through shared, participatory media events and through enabling playful encounters with strangers.

Exposure to strangers: shared and co-encounters

Although the internet and social media opens up a potentially universally accessible public sphere, a number of authors have argued that it contributes to 'networked individualism' (Wellman, 2001) where people gather around shared interests that equate more closely to private spaces than the public realm. This urban atomisation and fragmentation may actually promote the further withdrawal of urban social interaction from urban places into specialised electronic networks (Graham and Marvin, 1996, p. 232). A possible counter to these specialised, individual networks is the use of platforms that enable participation in public spaces through ad-hoc encounters with strangers. It is argued that the simple raised awareness of the presence of others in public space can contribute to the public realm (Sennett, 2010). When this exposure to strangers is coupled with a technological platform that enables these encounters to be shared or experienced in some form of co-present way, it can be argued that this can further promote the production of the public realm. Technological encounters in public space create the possibility for shared encounters; an experience I define as: 'the interaction between two people or within a group where a sense of performative, co-presence is experienced and which is characterised by a mutual recognition of spatial or social proximity' (Willis et al., 2008). This emerges in many different ways, but shared encounters tend to distinguish themselves by a one-off media event or encounter at a specific place, followed by a longer-term development of some form of shared presence spread across many places that is more about constructing relationships through encounters over time and space.

The first type of shared encounter is about constructing a background sense of awareness of strangers in public space that can counter the retreat into atomised social groups. This can happen when ubiquitous technologies such as situated interactive public displays in the city enable an inter-play between large displays and mobile end-user terminals that bring shared experiences into a public setting. These types of media offer a kind of shared viewing and common experience. According to Struppek (2006), such stages for the construction of publics can also be realised by connecting large outdoor screens with experiments in online worlds, the culture of collaborative content production and networking could be brought into a wider context. The liveness of an urban screens event is complimented by how they act as 'access points' (Hornecker, Marshall and Rogers, 2007; O'Hara, Glancy and Robertshaw, 2008). These projects document a role for urban screens within public spaces, by providing access or entry points for the staging of 'being together' with strangers. By creating legitimate 'access points' for strangers to be together in public, they allow for a temporal shared experience, and create a sense of background togetherness and shared involvement. An early example of how participation via the internet could create a dynamic, which challenges the creativity, of people in public space, is the project Blinkenlights that took place in Berlin in 2001 (Blinkenlights). The Chaos Computer Club, a German hacker community, used an empty building at Alexanderplatz (a central plaza in the former East Berlin) and transformed it into a giant pixel screen, by connecting the lights installed in the windows to a central computer system. Via a simple interface with the Internet, people could create their own animations and send them to the screen, or even play the computer game 'pong' on the screen. A special feature was 'love letters', self-made animations that were triggered via a mobile phone. The mobile phone became a remote control for engagement with the surrounding architecture. A number of recent projects have taken this one step further and explored the potential of connecting two remote locations live. For example, the Hole-in-Space project used the window metaphor in 1980 to create a link between New York and Century City, USA (Galloway and Rabinowitz, 1980). People who were walking past the Lincoln Center for the Performing Arts in New York City, and 'The Broadway', a department store located in the open air Shopping Center in Century City, LA, could suddenly see head-to-toe, life-sized, television images of people in the other locations giving the impression they were encountering each other on the same sidewalk. More recently the 'Screens in the Wild' project linked two UK cities through public displays and aimed to investigate how media screens located in urban space can be designed to benefit public life, rather than merely transmit commercial content. The project, which was initiated by researchers from The Bartlett at University College London and the Mixed Reality Lab at University of Nottingham linked public screens in London and Nottingham, and found that micro-publics of interest evolved, such as dog walkers (gen. Schieck et al., 2014).

The second type of shared encounter is where some form of mobile media is used to track or record images or locations in your everyday life enabling you to then share these images with others; forming a 'hybrid public'. Many social media applications have focused on the ability to show others what one is doing or seeing and to comment on each other's posts. The result is an expanded sense of observation of one another's lives, and a greater sense of 'knowing' each other across distance (Lewis, Pea and Rosen, 2010). Shifting the frame of interactivity from 'participation' in pre-established frameworks to collaboration and co-creation of new forms of interaction offers up new possibilities for publics. The 'co' or together model, rather than offering just a background 'awareness' is an important shift. For example projects such as Comob (Southern and Speed, 2009), create new relationships between people and places. Comob is a GPS enabled smartphone app that allows a group of people to see each other's geographical locations in 'real-time' and connecting them with lines. In studies they undertook, the creators of Comob found that it had a potential for use in the co-ordination of strategic spatial action, since each participant was able to see the rest of the group and co-ordinate their movements (Southern and Speed, 2013). On a larger scale, the use of mobile media to co-ordinate political movements ranging from Occupy to the Arab Spring has shown that publics can mobilise around common interests. But these mobilisations of people are not totally mobile or placeless; they still rely on the physical qualities of public space for the critical point of the encounter. The large, open public squares and parks such as Zuccotti Park, in Lower Manhattan, New York City, and Tahrir Square, in Downtown Cairo, Egypt, were crucial for both Occupy (which is described as 'capable of summoning an army with the posting of a tweet' [Feuer, 2012]), and the Arab Spring crowds were crucial for the gathering of likeminded people. Although people may gather and encounter others in social media, the point at which they come together to effect action still needs the sense of collective presence that is only really possible in an open, accessible public space.

Publics for intervention, interruption or interference

In 1960 Yves Klein jumped from the roof of '3 rue Gentil-Bernard' into the street below. The moment is captured in a photograph that shows Klein mid leap between building and pavement (The Metropolitan Museum of Art, n.n.) and was titled 'Leap into the Void'. Klein reproduced the image in a faux newspaper, Dimanche, with the caption: 'The Painter of Space Throws Himself into the Void!' Klein's intervention in a quiet suburban street is an early example of a media event. Klein's leap was an intervention; it disrupted the idea of the air as an empty space, and challenged his own mortality. Dayan and Katz (1992) claim that the 'media event' represents the privatisation of public space: they are events once experienced collectively in public space were increasingly consumed by greater numbers of people who watched from the privacy of their individual homes. McQuire (2006) contends that 'the 'media event' is in the process of returning to the public domain'. The media event contains the possibility for the construction of publics when it operates as a jolt or an intervention into a norm that causes reflection. Sennett argues that when individual experience is subjected to multiple collisions or jolts, these 'are necessary to a human being to give him that sense of tentativeness about his own beliefs which every civilised person must have' (1977, p. 296). This is about creating the conditions for the construction of an awareness and understanding of strangers in public space.

In public space it is important to remember the difference between such jolts and an 'intervention or interruption, and an unwanted interference' (Taylor, 2006). Many of the ways in which media intervene in public space are based around consumption; they are at best interruption and at worst interference. Surveillance and passive data gathering is one example where the city is brought into being through 'new software-sorted geographies' (Graham, 2005), or Shepard's 'sentient city' (2011) that seethe with our data. The possibility of external intervention is akin to the filtering, except that it's not the choice of the individual, but of the organisation holding your data. Recent media reports about how social network companies such as Facebook are not only using the monitoring of our location in public space to sell our data to private companies, but are also intervening to influence our behaviour. In an interview with Cameron Marlow of Facebook, who until 2013 led the Data Science team, he underlines the potential to interfere since 'Facebook News Feed is the thing that everyone sees and it controls how information is disseminated – it's controlling how information is revealed to society' (Simonite, 2012). By controlling, or filtering what we see, organisations can at best interfere and at worst intervene in our behaviour. In Klein's memorable photo the blanket held by his friends that broke his fall was erased to create the impression of the fall actually being a point between life and death. In our data driven world, companies can filter out or intervene with data that can similarly change our view of reality, and consequently influence our future behaviour.

Play, immersion and engagement

According to Sennett, play is a vital way for strangers to encounter one another in publics since 'playacting in the form of manners, conventions, and ritual gestures is the very stuff out of which public relations are formed' (1977, p. 29). The importance of playacting is that it involves testing out boundaries; it moves social rules from the background to the foreground of attention. It also creates one of the most permeable formats for strangers to encounter one another in some form of common stage, where they experience a sense of co-presence, shared goals and common outcomes.

Attention is the increasingly the measure we use to value engagement. In an 'attention economy' (Davenport and Beck, 2001) where you choose to commit your time and body, and whether you remember what you have seen or experienced is as important as where you choose to spend your money. But this all comes in degrees; there is no distinct on or off of attention, as the prevalence of multi-tasking in our everyday life shows. In fact much of the time we are busy shifting between one attention-seeking media to another; in public space we can be simultaneously moving between an email and a post on Facebook on a smartphone, to an advertisement in public space to the words of someone else's conversation. Goffman (1983) distinguishes how we manage attention as the way we move between foreground and background. To completely capture someone's attention is usually seen as an indication of a meaningful experience in public space; a meaningful moment where a relation between the person and the world is constructed. Everyone can remember a piece of music or theatre where the performance was so compelling that they became immersed in the event; often to such an extent that it suspended or altered his or her sense of reality. The experience can be highly individual but it can also be part of a sense of collective participation. An emerging form of immersive experience in urban space is applications and platforms that are based on what is termed 'augmented reality' (Sutherland, 1965). Augmented reality (AR) involves the overlaying of information onto a visual display that is indexed to the persons' actual location, activity or situation. Initially this was achieved with a cumbersome head-mounted display, more recently it is delivered through a smartphone screen or headsets such as Google Glass. The problem with AR is that it can reduce experience to spectating or 'viewers, seeing interfaces or graphical overlays' (Pedersen, 2009, p. 11). One of the key limitations to the construction of publics in augmented reality is that they require a screen that tends to need to be viewed from a singular viewpoint. Since the most recent models of AR, such as Google Glass are based on personalisation of experience (i.e. I only see what I want to see), there is little opportunity for collective experience, whether co-present or remotely, and even less for participation in something resembling a public space. Also, due to the expense and technical expertise needed to gain access to AR, currently it is creating private bubbles in public space or as one Scott Rigby, founder of AR company Immersyve, speculates; 'what will the consequences be of immersing yourself in a world that is isolated from the person standing next to you?' (Folger, 2014). If AR can find a way to bring backgrounds to the foreground, such as revealing backchat at a public event or facilitating collaborative actions, then it might engage with the performative possibilities of enhancing the public nature of gatherings, otherwise it will contribute to a privatisation of public spaces.

4.5   CASE STUDY: URBAN SCREENS

In this section I will use studies of real world examples to test out how the interaction with technology in urban public space might contribute to the public realm. The case studies involve a type of 'urban media' that has been termed 'urban screens', that are 'dynamic digital displays and visual interfaces located within urban public spaces' (International Urban Screens Association, n.n.). According to McQuire urban screens 'belong to a paradigm shift in the place of media technologies that is rapidly altering both the ambience and the dynamics of public space in contemporary cities' (2010, p. 568). These screens range from small-scale terminals in transit spaces, through to large scale screens located in central public spaces (Figure 4.2) through to the appropriation or installation of interactive lighting technology into or onto the facades of existing buildings. It also includes the use of smartphones and other mobile devices to interact with the fixed screens.

4.2 Football fans watching a match on the BBC Big Screen, Manchester, UK, during the 2006 Football World Cup (image courtesy of Sarah Griffiths, BBC Big Screens).

Square: being together

Large scale, fixed urban screens started to appear in public space in eighties, mainly as a result of the development of video wall technology, comprising of arrays of CRT screens arranged together to create a single surface. The first screens were primarily employed for advertising, and populated global, urban focal points such as Times Square in New York and Hachiko Crossing in Tokyo's Shibuya district. However in the noughties, the potential of urban screens to contribute to the public realm started to be explored. This included developments such as screens incorporated into the facades of buildings in central public spaces, such as Federation Square in Melbourne, Australia and BBC Big Screen; an installation of a series of large urban screens in UK cities. These screens focused on delivering content that would contribute to the public space, and did not rely on funding from advertising revenue. The UK Big Screens project started in 2002, when the UK public broadcaster, the BBC, installed a set of ten screens in central public spaces in ten British cities. They displayed mainly public broadcast content, including live events, but they also show a range of cultural content and public information and are integrated into site-specific events programmed by local partners such as city councils (McQuire, 2010, p. 572). The Big Screens project later expanded to twenty-one screens across UK, and aimed to broadcast localised content designed 'to bring people together for shared experiences, as well as encourage an interest in and conversation about local communities' (BBC, 2012). McQuire, in interviews with Bill Morris, director of BBC Live Events, records how the screen operated in two modes: 'event mode' and 'ambient mode'; where the screen split roughly between 'the 'event mode' of established crowd-pullers like live sport, where the screen is the pinnacle of attention, and 'ambient mode' where the audience tends to be more transient and distracted,' (McQuire, 2010, p. 573). In event mode the screens were used to broadcast key national sporting events, such as English national football matches, the Wimbledon tennis championships and the UK 2008 Olympic games. These events had the capability of capturing large live audiences; the Manchester screen had a capacity of 12,000 in an amphitheatre-shaped space in front of the screen and during the 2006 World Cup for the England-Paraguay match, more than 50,000 people watched the matches on big screens across the whole country (Morris and Gibson, 2006). During the football match, Mike Gibbons, chief project director of BBC Live Events, which programmed the screens, recalled "'there was this real feeling of why is there 8000 people in Victoria Square in Birmingham and 10,000 people in Manchester and 10,000 in Leeds all standing there in the pouring rain?'" (McQuire, 2010, p. 568). The scale of public viewing demonstrated a potential for public viewing to provide a platform for shared experience.

McQuire documents 'the emergence of large screens as a focal point for collective gatherings in public space' but with the difference that 'the screen does not so much substitute for a public gathering as become the occasion for one' (2010, p. 574). This can be shown in how the big screens in UK and Australia found an unexpected purpose as a focal point for collective acts of public mourning. For instance the BBC Big Screens emerged as a common platform for mourning around the time of the terrorist bombings in London in 2005. According to the BBC's Bill Morris; "'with the London bombings, people – not just in London, but in the other cities around the country – were gathering around the screens to watch what was going on ... When there was the three-minute silence that happened after the London bombings, maybe a week later, people gathered in quite large numbers at each of the screen sites to observe the silence'" (McQuire, 2010, p. 574). Morris observed that this was as a result of the fact that during significant, but unexpected national events people 'want to be with other people' in which the act of watching or participating together performs the connection of the individual to the wider community. At Federation Square in Melbourne, similar evidence of the public need to collectively mourn took place on 13 February 2008. On this date 8,000 people gathered to watch the newly elected Prime Minister Kevin Rudd in Canberra deliver a historic apology to Australia's indigenous stolen generations. According to Yue 'when the opposition leader's speech was broadcast, most in the crowd, like those on television, turned their back to the screen. Tears were shed and shared, as was the standing ovation at the end of the speech. As the apology turned to healing, musicians began to perform on the stage in front of the screen. Mobile phone messages sent by the crowd appeared on the screen: 'Our ancestors can finally rest in peace'; 'Sorry it's taken so long to say sorry'; 'Let's enjoy this day and think about those who have suffered from Australia's shameful past" (Yue, 2009, p. 270). In this format, the urban screen functions as an access point in the construction of a large-scale, public, shared experience.

Hybrid publics

As a site for intervention in the public sphere, activists are increasingly using multiple screen-based platforms to create mobile or hybrid publics. One example of this is how the Occupy movement has explored new models of hybrid publics; moving between discussion and debate in a physical public space and online discussions in a range of online spaces. In Massey and Snyder's account of the Occupy movement they recount how 'hybrid discussions were the norm for the working groups that handled the day-to-day and week-to-week activity of Occupy Wall Street … At the same time, another crowd assembled in a range of online spaces' (Massey and Snyder, 2012). They document how the hashtag #occupywallstreet, similarly operated as a form of public space so that 'the first of thousands of #Occupy hashtags enabled the spontaneous assembly of strangers on Twitter and other internet platforms' (Massey and Snyder, 2012). Urban screens also provided a presence in the park where 'note-takers projected their evolving documents on a screen in Liberty Plaza so that participants could respond to the minutes-in-the-making. Assembly meetings were live-streamed so that participants across the globe could follow in real time, and some were archived online in audio and video formats' (Massey and Snyder, 2012). The use of multiple platforms, linked up to a central physical space; in this case Liberty Square, meant that participation was not limited to those in the shared physical space or time, 'rather, there are spatial and temporal displacements, as media both extend the geographical reach of one's audience beyond the physical space and prolong the traces left by our actions in media archives and collective memory' (Iveson, 2009). In the series of protests, characterised as the Arab Spring, Tahrir Square in Egypt became a central place for activists to gather. However in a study of how the protestors mobilised using social media, almost half (48.2 per cent) had produced and disseminated video or pictures from political protest in the streets. Tufekci and Wilson found that although face-to-face contact was the primary way that people first heard about the protests in Tahrir Square, screen based media was used to disseminate videos and pictures from the political protests in the street. Almost half the protestors in the study had produced and disseminated video or pictures from political protest in the streets, and 'the leading platform for producing and disseminating visuals was Facebook, used by about fully a quarter of the sample (twenty-five per cent), and phones were a distant second, used by fifteen per cent (Tufekci and Wilson, 2012, p. 373). Similar patterns were found at Liberty Square, 'Facebook supported a weak form of political discussion that prefigured the stronger and more interactive deliberations that filled Liberty Plaza' (Massey and Snyder, 2012). If, according to Habermas, 'a portion of the public sphere comes into being in every conversation in which private individuals assemble to form a public body' (1974, p. 49), then the constellation of public space, social media and the dissemination of video recordings made in-situ constitute some of the conditions required to form a hybrid public sphere.

Streets: play

What is critical to remember about the success of these playful experiences is that they required significant planning, co-ordination and maintenance of the interaction in order for the public experience to be successful. One of the first experiments with how an urban public space could be transformed by playful interaction with an urban screen was Project Blinkenlights, described earlier in this chapter. In order to transform the façade of the 'Haus des Lehrers' building in Berlin into an interactive public display, the organisers installed over lamps behind the building's front windows in the upper eight floors of the building (Blinkenlights). The project required significant technical organisation. In just four weeks, 144 fluorescent lights were installed in the building's windows to create the eighteen by eight pixel screen, five kilometres of cable was laid and numerous relays and a central Linux server were specially set up to run without crashing for over five months. Similarly, the open format, actually involved significant editing and curation, according to Tim Pritlove, of CCC of 'over 800 films were submitted, only 250 messages were shown' (Krempl, 2002).

At the start of the chapter I described my experience of playing the augmented reality game 'Can you see me now?' Whilst playing it, it appeared as a seamless and immersive experience, and there was little awareness of the 'directors' or organisers involvement in the running of the game. But behind the scenes there was a team of 'stage managers' who worked to achieve this seamless experience for the players. According to a research-based study of the game from the perspective of how to create and manage the game experience, the creators point out that 'successfully staging a mixed reality game in which online players are chased through a virtual city by runners located in the real world requires extensive orchestration work' (Crabtree et al., 2004, p. 391). This 'orchestration work' included providing adequate support for monitoring and 'intervening in an event from behind the scenes In order to ensure a smooth experience for the participants' (Anastasi et al., 2002). In the case of the mixed-reality game, the management of the game was orchestrated from a control room, which operated on a number of platforms monitoring GPS accuracy, the location of participant positions and the monitoring of transmitted text messages. This extended to functionality that enabled the removal of particular players. In fact this orchestration also extended to some players, who worked out how to orchestrate the position of the runners so that they came into view of the public-play consoles. According to Benford et al. 'in both the Sheffield and Rotterdam experiences, the areas in which the public-play consoles were located contained small windows that looked out onto the physical game space. In both cases, some players reported enjoying deliberately positioning or moving their avatars in such a way as to cause runners to move into view. These rare moments of actually seeing a runner chasing their invisible avatar caused great excitement' (Benford et al., 2006, p. 108). Here, the players exploited the game format to try and 'see' the runners, and so literally played with the aim of the game 'Can You See Me Now?' What is interesting is the degree of production of the gameplay experience required to orchestrate a playful experience. Mixed reality games can create an interaction between passive spectator, active player and online screen in a way that suggests new forms of performativity in public space, based on a live experience mediated by a single screen and a group of players.

4.6   SUMMARY

The social experience of public space is characterised by two contrasting conditions; we are both present together with others, but also aware of them as strangers. The public realm is seen as being critical in the construction of publics or a sense of community that can counter the latter condition; the sense of alienation or strangeness that Simmel claims characterises the contemporary city. The public realm, typically identified by a city's squares, streets, theatres and cafés is altered by communications technologies and publics can now be performed in networked spaces as much as physically on the ground. In this chapter I highlight the emergence of 'networked publics'; that create opportunities for awareness and participation in the construction of publics, but in a mixed on-and offline space. This has implications for the way in which the built environment can operate as an open platform or 'stage' for public life, and suggests that interactions through social media and urban media also create a civic layer. In public space, the combination of online social networks and offline interactions can create the conditions for 'shared encounters' between strangers. In the case study discussion the example of urban screens is explored to understand further how fixed screens in urban space can create 'access points' for the staging of a sense of togetherness between strangers. The combination of online publics and urban screens is discussed in relation to the political movements of Occupy and the Arab Spring, where the link between a focal urban public space, such as Tahrir Square or Liberty Plaza, and interactions in social networks, such as Twitter has been shown to be critical for the organisation and enactment of political interventions. On a more playful level, the 'gamification' of encounters in public space, that can contribute to Sennett's concept of publicness through play, are made possible through location-based media and offer up the potential for playfully experiencing a sense of togetherness. But it must not be forgotten that these require considerable orchestration. New forms of publicness are emerging, that operate across the public space of the city and networked public spaces, but this requires thinking about the built environment of public space in terms of its capacity to provide a 'stage' for the enactment of publicness.

REFERENCES

Anastasi, R., Tandavanitj, N., Flintham, M., Crabtree, A., Adams, M., Row-Farr, J., Iddon, J., Benford, S., Hemmings, T. and Izadi, S. (2002). *Can you see me now? A citywide mixed-reality gaming experience.* http://bit.ly/1OKpU5G.

Andrews, J. and Taylor, J. (1982) *Architecture: A Performing Art.* Oxford: Oxford University Press.

Arendt, H. (1999) *The Human Condition.* Chicago: University of Chicago Press.

Arminen, I. and Weilenmann, A. (2009) 'Mobile presence and intimacy – Reshaping social actions in mobile contextual configuration'. *Journal of Pragmatics: An Interdisciplinary Journal of Language Studies*, 41 (10), pp. 1905–1923.

Auslander, P. (1999) *Liveness: Performance in Mediatized Culture.* London: Routledge.

Ballantyne, A. (2002) *Architecture: A Very Short Introduction*. Oxford and New York: Oxford University Press.

BBC (2012) *BBC update on: beyond the broadcast. BBC Outreach Newsletter*. http://downloads. bbc.co.uk/outreach/supplements/FocusOn_PublicPurpose2012.pdf (Accessed: 1 October 2014).

Benford, S., Crabtree, A., Flintham, M., Drozd, A., Anastasi, R., Paxton, M., Tandavanitj, N., Adams, M. and Row-Farr, J. (2006) 'Can You See Me Now?' *ACM Transactions on Computer-Human Interaction*, 13 (1), pp. 100–133.

Bertelsen, O.W., and Bodker, S. (2001) 'Cooperation in massively distributed information spaces', in Prinz, W., Jarke, M., Rogers, Y., Schmidt, K. and Wulf, V. (eds), *Proceedings of the Seventh European Conference on Computer Supported Cooperative Work*, Bonn, Germany. Dordrecht: Kluwer Academic Publishers, pp. 1–17.

Blast Theory *Our History & Approach*. www.blasttheory.co.uk/our-history-approach/ (Accessed: 11 October 2014).

Blast Theory *Can You See Me Now?: A game of chase played online and on the streets*. http:// www.blasttheory.co.uk/projects/can-you-see-me-now/ (Accessed: 1 January 2014).

Blinkenlights. http://blinkenlights.net/blinkenlights (Accessed: 15 August 2014).

boyd, d. (2010) 'Social Network Sites as Networked Publics: Affordances, Dynamics, and Implications', in Papacharissi, Z. (ed.), *A Networked Self: Identity, Community, and Culture on Social Network Sites*. Oxford and New York: Routledge, pp. 39–58.

Crabtree, A., Benford, S., Rodden, T., Greenhalgh, C., Flintham, M., Anastasi, R., Drozd, A., Adams, M., Row-Farr, J., Tandavanitj, N. and Steed, A. (2004) 'Orchestrating a mixed reality game "on the ground"'. *SIGCHI Conference on Human Factors in Computing Systems (CHI '04)*. Vienna, Austria: 24–29 April 2004 ACM, pp. 391–398.

Davenport, T.H. and Beck, J.C. (2001) *The Attention Economy: Understanding the New Currency of Business*. Boston, MA: Harvard Business School Press.

Dayan, D. and Katz, E. (1992) *Media Events: The Live Broadcasting of History*. Cambridge, MA: Harvard University Press.

Feuer, A. (2012) 'Occupy Sandy: A Movement Moves to Relief'. *The New York Times*. http:// www.nytimes.com/2012/11/11/nyregion/where-fema-fell-short-occupy-sandy-was-there.html?pagewanted=all&_r=0, 9 November 2012.

Folger, T. (2014). 'Revealed World'. *National Geographic*. http://ngm.nationalgeographic.com/ big-idea/14/augmented-reality-pg2.

Galloway, K. and S. Rabinowitz (1980). *Hole in Space*. http://www.ecafe.com/getty/HIS/ (Accessed: 1 October 2014).

Gehl, J. (1987) *Life Between Buildings: Using Public Space*. New York: Van Nostrand Reinhold.

Goffman, E. (1963) *Behaviour in Public Places: Notes on the Social Organization of Gatherings*. New York: Free Press.

Goffman, E. (1983) 'The Interaction Order'. *American Sociological Review*, 48 (1), pp. 1–17.

Graham, S. and Marvin, S. (1996) *Telecommunications and the City*. London: Routledge.

Graham, S.D.N. (2005) 'Software-sorted geographies'. *Progress in Human Geography*, 29 (5), pp. 562–580.

Habermas, J., Lennox, S. and Lennox, F. (1974) 'The Public Sphere: An Encyclopedia Article (1964)'. *New German Critique* (3), pp. 49–55.

Habermas, J. (1989). 'The Public Sphere: An Encyclopedia Article', in Bronner, S.E. and Kellner, D. (eds), *Critical Theory and Society: A Reader*. New York: Routledge, pp. 136–142.

Habermas, J. (1999) *The Structural Transformation of the Public Sphere: An Enquiry into a Category of Bourgeois Society*. Oxford: Polity.

Hampton, K., Livio, O. and Sessions, L. (2010) 'The Social Life of Wireless Urban Spaces: Internet Use, Social Networks, and the Public Realm'. *Journal of Communication*, 60 (4), pp. 701–722.

Harvey, D. (2006) 'The Political Economy of Public Space', in Low, S. and Smith, N. (eds), *The Politics of Public Space*. New York: Routledge.

Harvey, D. (2012) *Rebel Cities: From the Right to the City to the Urban Revolution*. London: Verso.

Höflich, J. (2005) 'The Mobile Phone and the Dynamic between Private and Public Communication: Results of an International Exploratory Study', in Glotz, P., Bertschi, S. and Locke, C. (eds), *Thumb Culture: The Meaning of Mobile Phones in Society*. Bielefeld: Transcript, pp. 123–136.

Hornecker, E., Marshall, P. and Rogers, Y. (2007) 'From entry to access: how shareability comes about'. *Proceedings of the 2007 conference on Designing pleasurable products and interfaces*. Helsinki, Finland: ACM, pp. 328–342.

International Urban Screens Association (n.n.) *about urban screens: about*. International Urban Screens Association. http://www.urbanscreensassoc.org/about/ (Accessed: 1 October 2014).

Ito, M. (2012) 'Introduction', in Varnelis, K. (ed.), *Networked Publics*. Cambridge, MA: The MIT Press.

Ito, M. and Okabe, D. (2005) 'Technosocial situations: emergent structuring of mobile e-mail use', in Ito, M., Okabe, D. and Matsuda, M. (eds), *Personal, Portable, Pedestrian: Mobile Phones in Japanese Life*. Cambridge, MA: The MIT Press.

Iveson, K. (2009) 'Too Public or Too Private? The Politics of Privacy in the Real-time City'. *Engaging Data Forum*. Cambridge, MA, 13 October 2009.

Jacobs, J. (2002) *The Death and Life of Great American Cities*. New York: Random House.

Kolarevic, B. and Malkawi, A.M. (2005) *Performative Architecture: Beyond Instrumentality*. New York: Spon Press.

Krempl, S. (2002). *Public Spaces Invaders*. http://www.heise.de/tp/artikel/11/11798/1.html, (Accessed: 1 October 2014).

Latour, B. and Yaneva, A. (2008) 'Give Me a Gun and I will Make All Buildings Move: An ANT's View of Architecture', in Geiser, R. (ed.), *Explorations in Architecture: Teaching, Design, Research*. Basel: Birkhäuser, pp. 80–89.

Lewis, S., Pea, R. and Rosen, J. (2010) 'Collaboration with Mobile Media: Shifting from "Participation" to "Co-creation"'. *6th IEEE International Conference on Wireless, Mobile and Ubiquitous Technologies in Education*. Kaohsiung, Taiwan IEEE Computer Society, pp. 112–116.

Licoppe, C. (2004) '"Connected" presence: the emergence of a new repertoire for managing social relationships in a changing communication technoscape'. *Environment and Planning D: Society and Space*, 22 (1), pp. 135–156.

Licoppe, C. (2010) 'The "Crisis of the Summons": A Transformation in the Pragmatics of "Notifications," from Phone Rings to Instant Messaging'. *The Information Society*, 26 (4), pp. 288–302.

Ling, R. (2005) *The Mobile Connection: The Cell Phone's Impact on Society*. San Francisco, CA and Oxford: Elsevier/Morgan Kaufmann.

Ling, R. and Yttri, B. (2002) 'Hyper-coordination via mobile phones in Norway', in Katz, J. and Aakhus, M. (eds), *Perpetual Contact: Mobile Communication, Private Talk, Public Performance*. Cambridge: Cambridge University Press, pp. 139–169.

Lofland, L.H. (1998) *The Public Realm: Exploring the City's Quintessential Social Territory*. Hawthorne, NY: Aldine de Gruyter.

Low, S. and Smith, N. (eds), (2006) *The Politics of Public Space*. New York: Routledge.

Manovich, L. (2006) 'The poetics of augmented space'. *Visual Communication*, 5 (2), pp. 219–240.

Massey, J. and Snyder, B. (2012) 'Occupying Wall Street: Places and Spaces of Political Action'. *Places Journal*.

McCullough, M. (2014) *Ambient Commons: Attention in the Age of Embodied Information*. Cambridge, MA: The MIT Press.

McQuire, S. (2006) 'The politics of public space in the media city'. *First Monday*, 4 (Special Issue #4: Urban Screens: Discovering the potential of outdoor screens for urban society).

McQuire, S. (2010) 'Rethinking media events – large screens, public space broadcasting and beyond'. *New Media & Society*, 12 (4), pp. 567–582.

Meyrowitz, J. (1985) *No Sense of Place: The Impact of Electronic Media on Social Behavior*. New York: Oxford University Press.

Mitchell, W.J. (2004) *Me++: The Cyborg Self and the Networked City*. Cambridge, MA: The MIT Press.

Moores, S. (2000) *Media and Everyday Life in Modern Society*. Edinburgh: Edinburgh University Press.

Morris, S. and Gibson, O. (2006) 'Blow to BBC image as Liverpool and London pull the plug on big screens'. *The Guardian*. http://www.theguardian.com/media/2006/jun/13/broadcasting.bbc1, Tuesday 13 June 2006.

Nye, D.E. (1990) *Electrifying America: Social Meanings of a New Technology*. Cambridge, MA: The MIT Press.

O'Hara, K., Glancy, M. and Robertshaw, S. (2008) 'Understanding collective play in an urban screen game'. *2008 ACM conference on Computer supported cooperative work (CSCW '08)*. New York: ACM, pp. 67–76.

Pedersen, I. (2009) 'Radiating Centers: Augmented Reality and Human-Centric Designs', *IEEE International Symposium on Mixed and Augmented Reality*. Orlando, FL, 19–22 October 2009, pp. 11–16.

Putnam, R.D. (2000) *Bowling Alone: The Collapse and Revival of American Community*. New York and London: Simon & Schuster.

Rheingold, H. (2003) *Smart Mobs: The Next Social Revolution*. New York: Basic Books.

Scannell, P. (1996) *Radio, Television and Modern Life: A Phenomenological Approach*. Oxford: Blackwell.

gen. Schieck, A.F., Schnädelbach, H., Motta, W., Behrens, M., North, S., Ye, L. and Kostopoulou, E. (2014) 'Screens in the Wild: Exploring the Potential of Networked Urban

Screens for Communities and Culture', in Gehring, S. (ed.), *The International Symposium on Pervasive Displays (PerDis '14)*. Copenhagen, Denmark June 3rd–June 4th 2014. ACM, pp. 166–167.

Schivelsbusch, W. (1995) *Disenchanted Night*. Berkeley, CA: University of California Press.

Sennett, R. (1977) *The Fall of Public Man: On the Social Psychology of Capitalism*. New York: Alfred A. Knopf.

Sennett, R. (2010) 'The Public Realm', in Bridge, G. and Watson, S. (eds), *The Blackwell City Reader*. Chichester, UK: Wiley-Blackwell, pp. 261–272.

Sheller, M. and Urry, J. (2000) 'The city and the car'. *Int. J. Urban Reg. Res.*, 24 (4), pp. 737–757.

Shepard, M. (ed.) (2011) *Sentient City: Ubiquitous Computing, Architecture, and the Future of Urban Space*. Cambridge, MA: The MIT Press.

Simmel, G. (1950) *The Sociology of Georg Simmel*. Trans., ed. and introduction by Wolff, K.H., Glencoe, IL: Free Press.

Simmel, G. (1971) *On Individuality and Social Forms: Selected Writings [of] Georg Simmel,* ed. Levine, D.N. Chicago and London: University of Chicago Press.

Simonite, T. (2012) 'What Facebook Knows'. *MIT Technology Review*.

Southern, J. and Speed, C. (2009) 'CoMob'. *International Symposium for Electronic Art*. Manchester: ISEA, p. 72.

Southern, J. and Speed, C. (2013) *Co-Mob Research*. http://www.comob.org.uk/?page_id=179 (Accessed: 1 October 2014).

Struppek, M. (2006) 'The social potential of Urban Screens'. *Visual Communication*, 5 (2), pp. 173–188.

Struppek, M. (2014) 'Urban Media Cultures Reflecting Modern City Development'. *Screen City Journal*, May 2014.

Sutherland, I.E. (1965) 'The Ultimate Display', *Congress of the Internation Federation of Information Processing IFIP*. New York City, 24–29 May 1965, pp. 506–508.

Taylor, K. (2006) 'Programming video art for urban screens in public space', *First Monday* Special Issue #4: Urban Screens: Discovering the potential of outdoor screens for urban society.

The Metropolitan Museum of Art (n.n.) *Leap into the void*. http://www.metmuseum.org/toah/works-of-art/1992.5112 (Accessed: 24 August 2014).

Thuma, A. (2011) 'Hannah Arendt, Agency, and the Public Space', Behrensen, M., Lee, L. and Tekelioglu, A.S. (eds), *Modernities Revisited*. Vienna IWM Junior Visiting Fellows Conferences.

Tufekci, Z. and Wilson, C. (2012) 'Social Media and the Decision to Participate in Political Protest: Observations From Tahrir Square'. *Journal of Communication*, 62 (2), pp. 363–379.

Varnelis, K. (ed.) (2012) *Networked Publics*. Cambridge, MA: The MIT Press.

Wellman, B. (2001) 'Little Boxes, Glocalization, and Networked Individualism', in Tanabe, M., van den Besselaar, P. and Ishida, T. (eds), *Second Kyoto Workshop on Digital Cities II, Computational and Sociological Approaches*. London: Springer-Verlag, pp. 10–25.

Whyte, W. (1988) *City: Rediscovering the Center*. New York: Doubleday.

Whyte, W.H. (1980) *The Social Life of Small Urban Space*. Washington, DC: The Conservation Foundation.

Williams, R. (1974) *Television, Technology and Cultural Form*. London: Fontana.

Willis, K., Roussos, G., Chorianopoulos, K. and Struppek, M. (2008) *Shared Encounters*. London and New York: Springer.

Yue, A. (2009) 'Urban Screens, spatial regeneration and cultural citizenship the embodied interaction of cultural participation', in McQuire, S., Martin, M. and Niederer, S. (eds), *The Urban Screens Reader*. Amsterdam: Institute of Network Cultures, pp. 261–278.

5

Times

5.1 DIGITAL RHYTHMS

Freeze

5.1 Freeze
flash mob,
Spitalerstrasse,
Hamburg, 2011.

Standing still in a busy street, whilst the ebb and flow of people continues on around you, can be a very revealing way of observing the city. Standing still with two thousand other strangers is a somewhat different experience. The flow of people stops, and just a few, slightly un-nerved shoppers, squeeze between the dense crowds of stationary pedestrians (Figure 5.1). This was the experience of a Flash Mob I participated in which took place on a sunny autumn Saturday in Hamburg's main shopping street; Spitalerstrasse in 2011. Flash mobs may now seem to be a bit passé and also perhaps, slightly pointless, but they do present an interesting experiment in the potential of the co-ordination of events through social media. The Flash Mob I took part in was first announced on a dedicated website and on a Facebook page less than a month before. Potential participants registered the likelihood of their participation, so that by the day of the event, eighteen hundred and sixteen people had indicated they would be 'attending', eighteen hundred and ninety eight as 'maybe attending' and four thousand two hundred and ninety six as 'not attending'. The event was scheduled for three o'clock in the afternoon, and in the preceding half an hour the street started to congeal. People clustered at the edges of the street, out of the flow of pedestrians, and you could notice the increasing density of people who were obviously not shopping or simply hanging out. At the appointed time a whistle blew somewhere along the street, and the street stopped. Some people adopted poses, some just stood still, but the transformation was fairly immediate. To be stood completely still in a public space with thousands of other strangers is very different to being in a crowd at a music or sports event, that is focused on some external activity. Instead, for that moment of time, the frozen crowd is the event, something that was reinforced by the number of non-participating bystanders who took out their mobile phones and started pushing through the frozen participants to film the scene. Three minutes can seem a long time; the lack of noise in the street was disconcerting but also highlighted the levels of background noise that we find normal in a typical street. After the three minutes, which seemed like at least fifteen, the distant whistle blew again, and the crowd dispersed as quickly as it had coalesced. Time restarted, the street moved again. Afterwards I met the organisers, who I had been communicating with via email and was expecting a slick group of media-savvy twenty-somethings. Instead I found myself having a lemonade with seven lively and unassuming teenagers, who were all still at school, and one thirty year old, all of whom had got into organising the flash mobs in their spare time after having learned about them on YouTube. We don't often have the opportunity to stop the flow of a street, to change the rhythm of its traffic. But social networks can enable people to co-ordinate and synchronise patterns of gathering and dispersion. They organise gatherings that are synchronised around times as well as spaces, and create disruptions to the dominant flows and rhythms of spaces.

Unfolding times

The city is not just a space, but it is also a temporal experience. It has rhythms and phases, routines and rush hours. The everyday life of the city unfolds not only in space, but also in time. Lynch highlights how 'the heart of our sense of time is the sense of 'now'. The spatial environment can strengthen and humanise this present image of time' (1988, p. 65). Lynch outlines how time is not homogenous, and that the experience of the passage of time has different qualities. He distinguishes between 'rhythmic separation – the heartbeat, breathing, sleeping and waking, hunger, waves, tides, clocks' and 'progressive and irreversible change – growth and decay, not recurrence and alteration' (Lynch, 1988, p. 65). Just as space is socially constructed so is time, and rather than 'a singular or uniform social time stretching across a uniform social space' (Thrift and May, 2001, p. 5), time is not only unevenly experienced it is also subject to different tempos, duration and rhythms. This is not a purely phenomenological reading, as it goes beyond the bodily experiences of the senses. According to Lefebvre (2004) in his study of 'rhythmanalysis', presence is of an innately temporal character and can never be captured by any representation of the present, such as people's movements, and natural processes but can only be grasped through the analysis of rhythms. 'For there to be rhythm, strong times and weak times, which return in accordance with a rule or law – long and short times, recurring in a recognisable way, stops, silences, blanks, resumptions and intervals in accordance with regularity, must appear in a movement' (2004, p. 78). Lefebvre contends that there is a rhythm 'everywhere there is interaction between a place, a time and an expenditure of energy' (Lefebvre, 2004, p. 15). Places and rhythms are thus mutually constituted by different times; slow, fast, rhythmic, banal, repetitive, cyclical, abrupt and endless.

Real-time city

Our built space is structured around time. In the past the clock tower was to many the central point of village life, as were seasons and local festivals. With industrialisation, the rise of global transport networks and the commodification of time resulted in space being co-ordinated and operated through timetables, arrivals and departures which created durations and rhythms through which space contained specific activities. The distance between office and home is often regulated by the commute; a journey that is measured in time, rather than distance. In fact the time of modernity has been characterised as a linear, sequenced, clock time where one thing follows another, days are made of hours and minutes and clocks co-ordinate the global passage of day and night. Lash and Urry argue that the ascendance of clock time contributes to the 'dis-embedding of time from social activities as it becomes significantly stripped of meaning; the breakdown of time into a larger number of small units; the emergence of the disciplinary power of time; the increasing timetabling and hence mathematisation of social life; and the emergence of a synchronised measure of life first across national territories and later across the globe with the development of Greenwich and "world time"' (1994, p. 229). Changing mobilities, globalisation and the ubiquity of network infrastructure are some of the key factors seen as culpable in the changing temporal patterns of the city. This is because in a network infrastructure, connections are made and broken, network fields are entered and left, encounters are planned often based around media use and availability. Castells, in his study of Network Society, argues that this creates a new spatio-temporal formation made of communication flows and their infrastructure (2007, p. 178). He uses this to make the, somewhat paradoxical argument that the space of flows creates a 'timeless time'. This is the 'de-sequencing of social action, either by compression of time or by the random ordering of the moments of the sequence; for instance in the blurring of the lifecycle under the conditions of flexible working patterns and increased reproductive choice' (Castells et al., 2007, p. 171). Others have referred to this phenomena as 'real time city' (Townsend, 2000), and Ling and Yttri introduce the 'softening of time' (2002, p. 163), where micro-coordination of meetings among urban nomads leads to a de-sequenced and ad-hoc connecting of times rather than a planned and agreed sequence.

If spatial structures and features are materialisations of temporal structures then this signals a reconfiguring of space as well as time. This works on a number of levels. On the one hand, it affects everyday times, for example where people use buildings to frame their commutes, journeys and movements in time. Buildings such as the clock towers and other distinctive visual buildings, that were once the landmarks for how people orient themselves, no longer play a role as co-ordinating how people meet and gather. On the other hand, there is time at the scale of history, where buildings change slowly although often on a long timeframe. Built structures start to change in different timescales; temporary, pop-up and throwaway are more representative models of new current models of time.

Uneven and non-linear times

We tend to associate spaces with times. The organisation of when we can 'use' a building is controlled by when it is open or closed. For many, office work has a particular time frame (the 9 to 5), time spent at 'home' also has specific times, and often there is a commuting time inbetween. Holidays are 'time-out', when regular time is suspended. Much of the way we think and reference time starts with a central datum of 'now'; it is linear and sequenced, with a before and after, and divisible in between; time 'stretches out in front of us' and can be planned and used. Lefebvre argues that we need to privilege other models of time; such as repetition of movements and action, linear and cyclical rhythms and phases of growth and decline to recognise the fact that 'everywhere where there is interaction between a place, a time, and an expenditure of energy, there is rhythm' (2004, p. 15). If spaces have times, it is not like a container, but a more contingent quality since 'every rhythm implies a relation of a time with a space, a localised time ... a temporalised place' (Lefebvre, 1991, p. 230). This also recognises that times and rhythms are different for different people, groups and countries, but that they become entangled since people 'repeatedly couple and uncouple their paths with other people's paths, institutions, technologies and physical surroundings' (Mels, 2004, p. 16). Anyone who has tried to co-ordinate a Skype chat with someone on a different time zone and with different schedules knows how messy and disjointed shared times can be.

Sending a letter meant a co-ordination with a transport infrastructure; the collection and delivery of communication had to be co-ordinated with planes and trains and distance to be traversed. Much communication today seems timeless and happens at no cost to time; email, texts, video calls, streaming and satnavs all work apparently instantaneously. They also create strange temporal disjunctions. Knorr-Cetina (2003) found that the networked connections of global financial markets are so closely co-ordinated that traders, in New York, Frankfurt and London, felt synchronised with the global market flows, and were 'disembedded' with their colleagues in their respective city offices. As places unfold over time, so time becomes a space. Increasingly the structures that underlie these times are also impacting on the way space is experienced. Online shopping has disrupted traditional shopping patterns, and work no longer happens in one office but moves between different meeting points, co-ordinated by laptops and mobile phones. In a complicated folding-in of these changing rhythms, the data gathered by social network companies is increasingly being 'data-mined' or 'crowd-sourced' as part of what Ratti has termed the 'senseable city' (McLaren, 2011). This decade is the first where large-scale data sets exist and are being used to understand how patterns of occupancy and behaviour can be used to effect real time planning of changes to behaviour. Examples include measuring air pollution in relation to car use, and subsequent proposals to limit car access to parts of the city when air pollution reaches certain levels (Schofield, 2014). This 'smart' city model also extends to transport, rubbish and other material flows of 'things' in the city, where a systems-based approach seeks to optimise rhythms and times of occupancy and use, filling gaps in time or spreading time out into more even distributions.

Aims

This chapter will explore the characteristics of the changing temporalities associated with the use of network technologies and also discuss what impacts these changing practices have on the temporal characteristics of the city and its infrastructure. It starts by discussing how time affects space, and opens up the temporal to consider socially constructed concepts of time such as rhythms and memory. The context of the chapter is the spaces of current temporal flows; airports and other transport nodes, and I explore how they take on different status when there is a shift from occupying spaces to inhabiting times or journeys. The focus of the case study is on the condition of the 'real-time city' where synchronous and networked data exchange opens up opportunities for people to work together in what has been termed the 'sharing economy'. There is also an exploration of how the use of airports is changing, highlighting how they are increasingly micro-cities, but with sleeping, working and shopping allocated to measured times, rather than functional spaces. Finally, the growing use of highly synchronised and real time social media exchanges make possible real-time pop-up spaces that exist for hours and maybe weeks. These have the possibility for reintroducing rhythms and patterns of decay that can counter homogenised global 'real time'.

5.2   TIMES AND RHYTHMS

Everyday rhythms

William Whyte studied the value of street life on the social life of the city. It was no mistake that he used the method of time-lapse film to observe what happened in the plazas of New York City. In studying the plazas, he found patterns of behaviour in time or rhythms of occupation, which he describes as follows 'In the morning occupation will be sporadic, a hot dog vendor setting up his cart at the corner, elderly pedestrians pausing for a rest, a delivery messenger or two ... Around noon, the main clientele starts to arrive, soon activity will be at its peak and stay there until 2pm ... In mid and late afternoon, use is again sporadic ... Ordinarily, however plazas are dead by 6.00 and stay that way until next morning' (Whyte, 1980, p. 18). This is not dissimilar to the view that Lefebvre had from the window of his Paris apartment overlooking rue de Rambuteau, and facing the Centre Pompidou where he observed 'Rhythms. Rhythms ... the music of a city' (2004, p. 36). Out of this he developed his study of 'rhythmanalysis'. Lefebvre distinguished between different types of rhythm; cyclical rhythms, which involve simple intervals of repetition, and alternating (or linear) rhythms. He highlighted how these rhythms are 'temporal elements that are thoroughly marked, accentuated, hence contrasting, even opposed like strong and weak times' (Lefebvre, 2004, p. 78), and they exist within 'An overall movement that takes with it all these elements (for example, the movement of a waltz, be it fast or slow)' (Lefebvre, 2004, p. 78). Lefebvre argues that media, such as TV and radio, segments and breaks up everyday rhythms and that the immediacy of such media effaces the experience of time as something that unfolds. In contrast to the cyclical 'tide' of the rhythm of the street he compares the mediatisation of the everyday as a 'swamp' (Lefebvre, 2004, p. 48). This argument proclaims that technologies create an abstract, globally consistent time. Edensor outlines how the automation of rhythms by measurement devices, texts and automatic door closers, along with other co-ordination tools such as diaries, alarm clocks, planners are about an organisation of time into manageable units (2010, p. 9). This affects not just how we organise our everyday schedules, but also how we plan journeys and meet people. The coordination of co-present activities via the technologies of travel, Green argues, 'requires greater attention or orientation to "clock time" as the prevailing organisation of temporally based activities, including attention to measurable, calculable, and linear units of standardised time' (2002, pp. 282–283). Clock time creates rhythms that mark morning, afternoon and night, but it is a universal measure, that cannot acknowledge differential rhythms. A long boring hour is the same measure of time as being at an event where time flashes by. In urban space, clock time creates a commonality of temporal reference, but it cannot capture the rhythms of a busy lunch hour, followed by sporadic occupation in an afternoon and an empty night.

As there is prime real estate, so there is prime time. But now extrapolate to an entirely asynchronous city. Temporal rhythm turns to white noise. According to Mitchell 'The distinction between live events and arbitrarily time-shifted replays becomes difficult or impossible to draw (as it often is now on the television news); anything can happen at any moment. When, for example, does an online forum take place, and where do you show up for it? You cannot say. The discussion unfolds over an indefinite period, among dispersed participants who log in and out at arbitrary moments, through uncoordinated posting and receipt of e-mail messages (1995, p. 16).

Many social media platforms (Facebook, Twitter, Instagram) use time as a structure to display, present and store a user's photos, messages. These timeline sequences are reverse-chronologically ordered, so all time starts with 'now' as the datum. For instance, a diary on a computer is always open at the current date, whereas locating 'today' in a paper diary requires searching amongst the pages. In these diaries, rhythms of activity will emerge depending on patterns of activity, but timelines flatten time out into sequenced stages of standardised measurements of days, hours, weeks and months. According to Crang, the increasing ubiquity of media means that we are no longer managing the patterns of our daily life, but instead become actors in the performance of mediated and non-mediated presence and communication. We don't choose how to make sense of time in our lives. In a study of media use in neighbourhoods in the North of England, UK, Crang et al. found that it was the mix of media across spaces and times that opened up possibilities since 'media use means that paths, interactions, and connections that people form are woven and remediated through intersecting arrays of new media' (2007, p. 2422). If media use starts to present a homogenous and co-ordinated timeline of everyday life then we will all start to operate on chronologically sequenced time paths, all with their own individual trajectories. It is where these trajectories cross that rhythms emerge.

Mobilities

The experience of time in the city is played out in the journeys, trips and movements we make. Patterns of movement over time reveal the nature of a person's presence in a space. On a broader scale, networks and mobilities are profoundly interlinked. Sheller and Urry underline that 'urbanism has always been associated with mobilites and their control, and continue to be so more than ever. The technologies, infrastructure, material fabric and representational machinery of cities support these mobilities, whilst also being shaped and re-shaped by them' (2006, p. 2). Because we're generally not very good at accessing content whilst we're moving, the passive and sedentary transit space of the commute is time we occupy to catch up on online activities; we go shopping, buy a book or just window shop. These patterns of movement are linked closely with the repetition we experience in our everyday lives. Contrary to what we may imagine of the diversity in our lives, they are in fact extremely repetitive in terms of where we go. We tend to have work, school and travel commitments that require us to arrive, stay and leave to and from a range of places in a fairly recurrent sequence. This is highlighted by the dominance of the daily commute; we tend to take the same routes between these places and this repetitive structure actually makes our lives less complicated. We don't have to figure out which way to go afresh each day; we simply repeat the sequence.

Memory, remembering and forgetting

Increasingly social media is starting to tap into to our personal histories, and not just focus on the temporal here and now. The Internet has created a permanent record that memorialises everything you've ever said, done or experienced, that has been characterised as 'the end of forgetting' (Rosen, 2010). Whilst memory and narrative are starting to be given new status in the 'now' time of social media, the automated and unedited recording of a collective memory is problematic. In a society in which everything is recorded, Mayer-Schonberger notes that our 'digital representations will forever tether us to all our past actions, making it impossible, in practice, to escape them … without some form of forgetting, forgiving becomes a difficult undertaking … memory impedes change' (2009, p. 125). Hill points out that the photography app, Instagram, has created a series of filters such as '1976' that 'transform the look and feel of the shot into a memory to keep around forever' (Hill, 2012), so that they disassociate the image from when it was actually taken, it is a false memory, a piece of nostalgia.

Facebook has developed a way in which to create a sense of history through its timeline feature, which aims to 'represent your history in a way that mirrors personal memory. The most recent section are (sic) your freshest memories and are all apparent. As you travel back in time the years become abbreviated and only the highlights are initially visible. For users who have filled out time periods before they joined Facebook, it's witness to what they or others remember about their past and starts to form a collective memory as classmates post their 2nd grade photos or parents tag their children' (Hill, 2012). With this feature there lies the problematic that a recorded everyday history cannot be forgotten, and also another challenge is how to recognise when a human passing does not automatically mean that a Facebook page dies. In creating a platform that will always remember, there is also the problem that it never forgets.

5.3   HISTORICAL CONTEXT

Technological times: from clocks to GPS

Up until the end of the nineteenth century, scheduled and co-ordinated time was not the norm; there existed no mechanism to co-ordinate a universal time across separate geographical locations. Indeed one of the primary challenges of seafaring in the sixteenth and seventeenth centuries was to find a way to map the navigational charts of the starts to some form of temporal datum in order to be able to work out where a moving ship was on a cartographical map. With the advent of industrialisation, there was an imperative to standardise time for co-ordination and consistency in industrial production. The railways were the first to introduce a standard time towards the end of the nineteenth century, although it took a number of years for it to be fully adopted. More generally, the scheduling of transport systems at the end of the nineteenth century, together with the rise of mass circulation newspapers, the advent of the telegraph and telephone, radio and television contributed to a uniformity of measurement and co-ordination of time. Prior to this different geographical regions operated on different time zones. For example, Kern notes that in 1883 more than two hundred local times were encountered by a traveller on a railway journey from Washington to San Francisco (1983, p. 12). Harvey points out that 'it was only through the conquest of space after 1840 that an abstract, objective and universal sense of time came to dominate social life and practice' (1989, p. 175). These tightly scheduled times profoundly changed the rhythm and form of urban life in the shift from the nineteenth to the twentieth century. The introduction of the atomic clock added a level of accuracy to the measurement and co-ordination of time. Over the last twenty years the introduction of global constant time has added a global time constant, which is used by almost all navigational devices and many computers, that creates an accuracy of time that means there is no deviation. This relies on a service called the Network Time Protocol (NTP), which checks a computers' time against a more accurate server, such as an atomic clock (Pascoe, 2011). Similarly Global Positioning Systems (GPS) use a network of twenty-four orbiting satellites to create a global time constant that enables any position on the earth to synchronise time. Each GPS satellite contains multiple atomic clocks that contribute very precise time data to the GPS signals. GPS receivers, such as satnavs and smartphones decode these signals, effectively synchronising each receiver to the atomic clocks, that enables users to determine the time to within 100 billionths of a second, without the cost of owning and operating atomic clocks. This degree of accuracy is now crucial to communication systems, electrical power grids, and financial networks, which all rely on precision timing for synchronisation and operational efficiency. For example, wireless telephone and data networks use GPS time to keep all of their base stations in perfect synchronisation. This allows mobile handsets to share limited radio spectrum more efficiently. Similarly, digital broadcast radio services use GPS time to ensure that the bits from all radio stations arrive at receivers in lockstep. This allows listeners to tune between stations with a minimum of delay (*Timing*, 2014). The accuracy of GPS means that we are able to use mobile and navigational devices to accurately map our location, but it also means that any sense of rhythm or seasonal time can only be measured in relation to global time.

Social times: from journeys to global commuting

Just as uniformity in the measurement of time allowed the possibility of navigating safely to new territories, it also created structures for dividing the distance between places into separations of time. Simmel first noted at the turn of the twentieth century that 'spatial separation results in the making all waiting and breaking of appointments an ill-afforded waste of time. The techniques of metropolitan life is not conceivable without all of its activities and reciprocal relationships being organised and coordinated in the most punctual way into firm, fixed framework of time which transcends all subjective elements' (1971, p. 328). This had a profound effect on the organisation of urban space, since buildings and their associated activities are located in relation to the travel time and accessibility of one location to the next. In the late nineteenth century Marchetti (1994) developed a constant for measuring the average amount of time spent commuting each day, which was approximately one hour. This hypothesises that although forms of urban planning and transport may change, and although some live in villages and others in cities, people gradually adjust their lives to their conditions (including location of their homes relative to their workplace) such that the average travel time stays approximately constant (Marchetti, 1994). Over the last fifty years this has been affected not by the growing speed of car and air travel, but also by the airplane. The modern airplane, introduced in the 1950s, facilitated longer, transcontinental flights, and from 1975 to 2010, the number of scheduled aircraft departures in the United States more than doubled, from 4.5 million to 9.3 million (Lyster, 2013). Time and mobility are closely linked, and with the reduction in travel time enabled by modern forms of transport we are becoming more and more mobile. A US Federal Aviation Administration report predicts that domestic air travel will nearly double in the next two decades, reaching 1.2 billion annual passengers by 2032 (Ferdinando, 2012). The rise in internet use in the last twenty years has created a 'real time' where activities and events are co-ordinated through network connections that synchronise social connections almost instantaneously.

5.4   DIGITAL RHYTHMS

Although not defined by one type of technology or interaction, the changing temporal effects of technology impact in numerous ways. In this section I look at how technology, and particularly highly synchronised interactions enabled through social networks create new ways of occupying times, rather than spaces. Additionally, the hyper-connected nature of 'real-time' activities creates newly differentiated urban rhythms; some of this involves an individual slowing down or 'toggling' off, whilst for groups it allows for highly co-ordinated temporal events or 'swarms' that are experienced collectively.

Pop-ups and swarms

Towsend has highlighted how mobile media creates a commodification of time, so that 'time becomes a commodity that is bought, sold, and traded over the phone. The old schedule of minutes, hours, days, and weeks becomes shattered into a constant stream of negotiations, reconfigurations, and rescheduling. One can be interrupted or interrupt friends and colleagues at any time' (Townsend, 2000, p. 9). This type of live co-ordination also contributes to the ability to mobilise large groups or 'swarms'. Swarming is a technique developed from military scenarios where it is successful at mobilising a myriad of small, dispersed, networked clusters. It appears unstructured, but it is, in fact, deliberately organised and coordinated through network connections (Arquilla and Ronfeldt, 2000). More recently swarming has emerged through social media use and is characterised by what Rheingold terms 'thumb tribes' (2003, p. 1). 'Thumb tribes' are behind the way that flash mobs manage to mobilise and co-ordinate a group of strangers at specific locations and times. Distributed by ubiquitous media such as Flickr, Twitter and Facebook, flash mobs and urban swarms create temporary, situated user-generated scenographic practices as an embodiment of what Wasik provocatively calls 'viral culture' (Wasik, 2010). It enables an event to be organised through fast dissemination on social media and for it to develop participation without as yet developing an agenda. This means that the flash mob moves beyond ideas of participation and entertainment in an interplay between online and real life (Brejzek, 2010).

The flipside of swarming and the mobilisation of groups of individuals into collective action is that commercial organisations can similarly tap in to the behaviour patterns of individuals and develop marketing that influences or targets certain types of individual based on these patterns. Current location-based mobile applications can make sense of complex social systems by recording and visualising daily rhythms of behaviour through 'reality mining' (Eagle and Pentland, 2006). This process gathers data on people's movement patterns to reveal where certain types of people congregate and when. This type of data-analysis using complex algorithms may show, for example that a particular demographic heads to bars in the city centre between 6 and 9 pm on weekdays. The traces left by people and groups can this be mapped onto the urban space over time revealing a pattern of movement and presence that can be viewed as an aggregate, such as in Real Time Rome by MIT Senseable City Lab (MIT SENSEable City Lab, 2006). These mapped patterns of mobile phone activity onto the city map, reveal sites of intense activity as well as revealing changing temporal patterns of mobile phone use. With the growing importance placed on social networks and personal recommendation systems these patterns of presence will not only reveal behaviour, but also increasingly be used to influence it. This will have a resultant effect on how people choose to move and act, so places revealed to be trending spots for certain social groups may become busier at certain times, whereas real-time information on traffic jams encourage others to take alternative routes. The city becomes a live swarming system, adjusting or intensifying its flows in real-time.

Modulating the rhythms: slowing down and toggling

Daydreaming is one way to be somewhere else, to drift away. It is a favourite activity on long journeys and in boring lectures; a way of occupying 'dead' time. But our daydreaming is turning into media time. We're becoming dependent on live media, updates and messages to fill in apparently 'dead time'. According to Zimbardo 'technology creates a funny kind of obsession with time, but it's this very short-focused, immediate-present time. We're simply being in that moment to take the next action' (Gregoire, 2013). Studies show that, globally, on average one in five people checks a smartphone for email, text and social media updates at least every ten minutes, and in the US, two out of five people check at least once every ten minutes (Cisco). Our attention is almost constantly toggling between one activity and another. This means we seek to harmonise the discordant rhythms by bringing various aspects of the event to the foreground and letting others remain in the background. This may be active decisions (time ordered), or may be passive (time measured) or even just an awareness (felt time). For example, a mobile phone ringtone or text notification demands a response regardless of the appropriateness of the location or availability of the recipient, creating a state which Licoppe (2010) refers to as the 'crisis of the summons'. People have learnt to deal with the demands of such media by choosing where to focus their attention by switching their sensory and mental attention between the media device or screen and the features of the real world. In this way a person toggles their involvement from the physical space to media space like an on/off type effect. This has an impact on how people move in the space since rather than simply moving faster or slower the rhythm of motion starts and stops, as the person's interaction with media switches between attention to the media and attention to the physical environment. The person may choose when to toggle, such as deciding to make a phone call or send a message and seek out a suitable space in the environment to do this. Often this means moving to a place out of the dynamic flow of pedestrian traffic; a corner or niche to stand in or a bench to sit on. But the media use may be asynchronous with the space, such as when someone receives an SMS message or an LBS alert about a location. The toggling is then asynchronous with the rhythm of the space they are in; it adds a secondary, non-spatially defined temporality to the space that is often at odds with the rhythms in the physical place. Thus phenomena are observed where people literally bump into things or people, where they suddenly stop in the pedestrian flow of a crowd. As well creating different rhythms, time is slowing down in public. We spend more time lingering, working and meeting in public spaces and transit spaces. Rather than wasting time in inbetween spaces, we use it on catching up with shopping, chat and work. Airports, stations, harbours, and parking lots and even Wi-Fi enabled parks, are less and less seen as spaces for time wasting or daydreaming. They are increasingly spaces to be filled for time to be occupied by interaction with phones, tablets and other media, and because we are more occupied during this time, the spaces become more static in character, they slow down, even if they are literally moving.

Past, present and narratives

Time is also revealed in a pattern of past experiences. Many places are embedded with the history of past events and encounters, and these histories frame how we approach it and the way in which the movements and rhythms of human and non-human activity are registered in lived space. The history of space thus begins with the spatio-temporal rhythms of nature as transformed by a 'social practice, imposing the 'meshwork' of mental and social activity upon nature's space' (Lefebvre, 1991, p. 117). Location-based media has allowed the use of narrative to allow for immaterial, hidden, misunderstood and changeful qualities of space to be made material through location- based oral and textual narratives. Hight (2013) asserts that 'we can write with the physical world, and we can allow it and past moments to again have voice'. Using the metaphor of archaeology, he describes how locative media can be used to convey layers of information, stories, moments, people, lost things and past in places. Hight uses the example of his own co-authored project 34 North 118 West to reveal hidden histories in the urban landscape, using audio and first person historical narratives to enable what Hight refers to as a 'conversation' between place, its infrastructure, the movements of the person and digital information. Hight counters that this approach can allow people to convey their layers of information, stories, moments, people, lost things and past(s), and that this can work as a strategy against the augmented layering of space with abstract information.

5.5   CASE STUDY: 'REAL TIME' CITIES

Over the last twenty years a condition termed 'the real time city' has emerged that is characterised by a highly synchronised set of connections that creates a technology-driven set of rhythms in the city. These digital rhythms increasingly shape many everyday activities and spaces. In this case study I look at the impact of 'real-time' on temporal spaces; those of journeys, transit spaces and destinations, with a focus on airports and roads as sites of intensified temporal and technological connectivity. I explore three facets of this condition. Firstly, the way that airports are a particular type of space that have started to become organised around time rather than space. Secondly, how the emerging 'sharing economy' of hyper co-ordinated taxi rides and room-sharing apps and social network platforms creates a new currency of time. Finally I look at how the proliferation of pop-up, ad-hoc and temporal events within more fixed spaces re-introduces new rhythms into 'timeless' spaces of airports and other transit zones.

Real time cities on the move

Commuting or traveling was traditionally seen as unproductive time, it was often referred to as 'time-consuming'. Yet the increased use of mobile connectivity means that travel time has become one of the most connected and communicative times of a day. In a 2004 US study of travel behaviour, data showed that people made on average four trips a day and spent eighty minutes on the move (Bose and Sharp, 2005). Almost ninety per cent of these travel times were linked to a commute. Commuting or travel time is increasingly used for interaction with online information via a smartphone or mobile phone (Haddon et al., 2003). The length of a commute and the practicality of accessing mobile media or the Internet during the commute are thus time linked. This is recognised by the big transport companies, and even being underground is no longer a barrier. In London, UK, a free Wi-Fi is available to users of most of the major mobile networks across 120 London Underground stations, while mobile provider, EE has a deal to provide 4G connectivity in the Channel Tunnel between England and France in 2014. In a 2013 survey, ninety per cent of UK consumers reported that they browsed and shopped on smartphones and tablets during their commute (Macleod, 2013). The interesting thing about being mobile with a mobile Internet connection is that you can undertake activities in transit that used to involve travel. For instance, in a survey of mobile Internet activities it was found that '59 per cent of those surveyed have taken to commuter shopping because of the convenience, believing that it saves them time in the long run. Music and books (23 per cent), clothes (22 per cent) and food (17 per cent) were the most popular purchases whilst commuting, but travel was the most popular category to research, with a third looking up holidays' (Macleod, 2013). So whilst we're mobile we think about other journeys, and use the internet to plan them, in a manner referred to by Sheller and Urry (2000) as 'dwelling in motion'.

5.2 Passengers waiting for flight information fill cubicles throughout Atlanta Hartsfield-Jackson International Airport on Jan. 10, 2011, in Atlanta, Georgia (copyright Getty Images/ Jessica McGowan).

The airport is one of the physical realisations of the real-time city, because it is a transit space that acts as a hub of connectivity in terms of time rather than space. The world's largest airport; Hartsfield-Jackson Atlanta International Airport has, on an average day more than 250,000 travellers pass through its terminals which accounts for 95 million passengers annually (Figure 5.2). This represents a small city constantly being created and dispersed every single day, with an ever-changing population. Atlanta is so busy because it is the most centrally connected of all American airports in terms of flight time; it is within a two-hour flight of eighty per cent of the population of the United States (Ewalt, 2013). Atlanta may be the biggest, but the fastest growing is Dubai International Airport, with an annual flow of 60 million passengers, despite the local population only being 168,000 people. Dubai's international connections have grown so quickly in large part because of its location: the airport is within an eight-hour flight of two-thirds of the world's population (Mawoud, 2014). Dubai is successful because it offers the shortest connections in terms of time to global destinations; it connects a space of journeys to and from it rather than being a destination or departure point itself.

The increasing use of downtime to be online in airports is matched by the provision of connectivity via Wi-Fi and other sources. In the US, seventy nine per cent of airports offer free unlimited Wi-Fi, whilst many international airports increasingly offer charging stations. This reflects their passengers needs and the technology that accompanies them on their journey, since as many as seventy per cent of airport travellers will travel with a smartphone or tablet (*Mobile in the Airtravel Industry Report 2014*, 2014) compared with sixty per cent smartphone ownership for the average population (Pew Internet Project, 2014). Hartsfield-Jackson airport has installed over 1377 power outlets, an average 8.1 per gate freestanding charging stations, and many of these power outlets are in the form of a charging station comprises of three lots of two-plug outlets and two USB ports (Sullivan, 2011). The airport also installed workspaces for working travellers, comprising clusters of 4, 8, 10, and 12 'recharge stations', that are cubicle desks with power outlets at 19 locations, which is 240 powered workspaces (Sullivan, 2011). By providing these working and charging spaces within the space of a commute, the space is no longer 'down-time' but highly connected real-time work and shopping spaces.

Real-time coordination

One of the ways that flows of mobile travellers are being transformed through technology is in the way they access hyper co-ordinated and highly connected real-time services. The temporal micro co-ordination made possible through hyper-connected smartphone and internet platforms is being rolled out to many everyday services; the delivery of take-away food (for example seventy per cent of all Domino's Pizza takeaway orders are placed online [Robinson and Davies]), the matching of household tasks to people in the local community (TaskRabbit), the transformation of spare rooms or sofa's into hotel rooms (AirBnB) and sharing bike hire (Getaround). Exploiting the model of swarming, the new tools of the 'sharing economy' (*The Economist*, 2013) such as Uber, Hailo and Lyft work by co-ordinating times of people in their cars and taxis to those wanting to travel. Uber operates a global network of town cars, and a mobile interface that allows any customer to request a car pickup. Customers are kept up to date with text messages as the status of their request changes; when a driver accepts the request, when the driver is less than a minute away, and if the ride has been cancelled for any reason. The feedback is also co-ordinated, so that 'real-time passenger feedback means that drivers who consistently receive low ratings can be dropped from the service' (Goldwyn, 2014). These services work by micro-coordinating spare time and space. Hailo, a mobile taxi service found that drivers spend forty to sixty per cent of their shifts with empty cabs, and Jay Bregman, Hailo co-founder reports the potential this offers since; 'the transit market … have massive fifty percent plus inefficiency, and these problems could not have been solved before today' (Jeffries, 2013). This focus on optimising time is underlined by the fact that the SMS feedback system used by Uber works on the basis that 'fast delivery of customer notifications is critical when the transaction is in real-time. A delivery delay of more than a minute can leave a customer standing in the rain waiting for a ride they didn't know was cancelled' (*Twilio*, 2014). With co-ordination moving in to timescales of a minute, these platforms create new timescales of occupation and travel, and also shift ideas of temporal ownership. Instead of needing to own a car that lies unused for a good proportion of the time, you can now purchase the use of a car for a time when it is needed.

Pop-up spaces

The shift from sequenced, linear time structures to real-time also allows for the re-introduction of cyclical, sporadic and temporal events and experiences. In the context of airports, this has meant that they have now started to offer spaces based entirely on temporal activities and inhabitation. Atlanta airport offers the service; 'Minute Suites', that are spaces within the airport created for travellers to 'nap, relax or work'. A 'Minute Suite' is a seven by eight foot space (*Minute Suites*) that is sound isolated, and supplied with a TV and high speed Wi-Fi. The spaces are charged by the minute (hence the 'Minute' Suite name), and were used by over 18,000 people in 2013 (*Minute Suites*). Meanwhile London Heathrow airport has tried to create a temporal pop-up park and picnic site in what they call a 'sensory park pod', even 'pumping in the smell of freshly cut grass along with the sounds of birdies and other critters enjoying themselves'. As part of the service passengers are able to choose their airline meal from a range of suppliers and create an 'on-board picnic' for their upcoming flight' (Burns, 2014). Other attempts to re-introduce a sense of natural time into real-time spaces have included the hosting of seasonal pop events. This includes offering 'frozen yogurt in the summer, artisan chocolate at Easter, flip-flops and sandals during the summer' (Baskias, 2014). In Copenhagen a similar pop-up space in the form of a restaurant has been created. The 'Hallo Hello' pop-up restaurant offered travellers the chance to enjoy a three-course meal with a stranger. According to the creators 'the objective is to establish a dialogue between two travellers who would otherwise never have talked with each other. We do so by offering them a dining experience where travellers who do not know each other meet and share a meal. Hopefully, they will take this mood along to their flight and when they travel into the world' (Copenhagen Airports, 2014). The experience was intended to continue beyond the airport and into the plane journey, and the dessert course of the menu was served in a goodie-bag which the guest could take along and share on their flight or train ride (Copenhagen Airports, 2014). Pop-up and event spaces work in a social media connected world, because events can be publicised and created in short timescales, and the large numbers of commuting passengers with time to spare create a captive audience. The pop-up space works because it creates a time-based experience or 'atmosphere' that doesn't require costly infrastructure. The introduction of such pop-up events is one solution to counter the homogenisation and de-sequencing of time from natural rhythms in real-time cities.

5.6   SUMMARY

The temporal dimension is often overlooked in the discussion around the effects of networked infrastructures on urban space. Where it is acknowledged, much of the focus is on the homogenisation and stripping time of its context; into what Castells et al. characterise as 'timeless time' (2007, p. 171). In this chapter it is argued that interactions through networked media and infrastructures also contribute to changing rhythms and sequencing of everyday activities. This is tightly linked with changing mobility patterns; both through global connections and at a more local level through changing commutes and shopping. In online social networks a sense of time is becoming much more focussed on the now, with features such as timelines and highly synchronised updates that have implications for how we forget events or people and make sense of the past. Changing digital rhythms also contribute to new patterns of social gathering and activism, such as swarming and flash mobs. These are bound up with what has been termed the 'real-time city', and one of the many implications of this condition is the increasing issues with how our temporal behaviours become revealed through our media use, and our activities in time and space consequently become both commodified and also shareable. The real-time city is explored through the lens of a case study looking at spaces of transport such as airports, cars and trains. The ever-increasing number of plane journeys annually has had implications for airport spaces, and it was shown that passengers tend to occupy them as temporal rather than spatial units. This is demonstrated through the emergence of rentable sleep units, which can be paid for by the minute. The co-ordination of transport activities also underpins many of the practices of the new 'sharing economy', with taxi rides, bike hire and room rental now happening in real-time. Finally the broader emergence of pop-up spaces can also be seen as being driven by highly mobile and socially networked 'smart mobs' that introduce new digital rhythms of occupation and activities in the city.

REFERENCES

AirBnB. *How it Works*. https://www.airbnb.co.uk/help/getting-started/how-it-works (Accessed: 1 October 2014).

Arquilla, J. and Ronfeldt, D. (2000) *Swarming and the future of conflict*. http://www.rand.org/pubs/documented_briefings/2005/RAND_DB311.pdf: RAND Corporation.

Baskias, H. (2014) 'Pop-up shops spice up airport retail options'. *USA Today*. http://www.usatoday.com/story/travel/flights/2014/02/19/airport-pop-up-stores-retail-shopping/5581917/ (Accessed: 19 February 2014).

Bose, J. and Sharp, J. (2005) 'Measurement of Travel Behavior in a Trip-Based Survey Versus a Time Use Survey: A Comparative Analysis of Travel Estimates Using the 2001 National Household Travel Survey and the 2003 American Time Use Survey'. *ATUS Early Results Conference*. Bethesda, Maryland: 8 and 9 December 2005.

Brejzek, T. (2010) 'From social network to urban intervention: On the scenographies of flash mobs and urban swarms'. *International Journal of Performance Arts and Digital Media*, 6 (1), pp. 109–122.

Burns, J. (2014) 'Heathrow Installs Sensory Park Pod in T2'. *Airport World*.

Castells, M., Fernández-Ardèvol, M., Qiu, J.L. and Sey, A. (2007) *Mobile Communication and Society: A Global Perspective*. Cambridge and London: The MIT Press.

Cisco *Toothpaste, Toilet Paper, and Texting– Say Good Morning to Gen Y*. http://newsroom.cisco.com/press-release-content?articleId=1114955 (Accessed: 7 August 2014).

Copenhagen Airports (2014) *Share your food with the neighbour – a social experiment at Copenhagen Airports*. https://www.cph.dk/en/about-cph/press/news/del-din-mad-med-naboen---socialt-eksperiment-i-kobenhavns-lufthavn-/ (Accessed: 3 October 2014).

Crang, M., Crosbie, T. and Graham, S. (2007) 'Technology, timespace and the remediation of neighbourhood life'. *Environment and Planning A*, 39 (10), pp. 2405–2422.

Eagle, N. and Pentland, A. (2006) 'Reality mining: sensing complex social systems'. *Personal and Ubiquitous Computing*, 10 (4), pp. 255–268.

Edensor, T. (ed.) (2010) *Geographies of Rhythm*. Aldershot: Ashgate.

Ewalt, D.M. (2013) 'America's Most Crowded Airports'. *Forbes*.

Ferdinando, L. (2012) *FAA Expects Travel on US Airlines to Nearly Double in 20 Years,* Voice of America. 7 March 2012. Available at: http://www.voanews.com/content/faa-expects-travel-on-us-airlines-to-nearly-double-in-20-years-141988063/179138.html.

Getaround *How it Works*. https://www.getaround.com/tour (Accessed: 1 January 2014).

Goldwyn, E. (2014) 'Will Uber Destroy the Driving Profession?' *The New Yorker*. http://www.newyorker.com/tech/elements/will-uber-destroy-the-driving-profession (Accessed: 8 August 2014).

Green, N. (2002) 'On the move: Technology, mobility, and the mediation of social time and space'. *Inf. Soc.*, 18 (4), pp. 281–292.

Gregoire, C. (2013) *How Technology Speeds Up Time (And How To Slow It Down Again). The Huffington Post*. 12/06/2013. Available at: http://www.huffingtonpost.com/2013/12/06/technology-time-perception_n_4378010.html.

Haddon, L., de Gournay, C., Lohan, M., Östlund, B., Palombini, I. and Sapio, B. (2003) *From Mobile to Mobility: The Consumption of ICTs and. Mobility in Everyday Life* http://www.lse.ac.uk/media@lse/whosWho/.../Mobility and ICTs.pdf.

Harvey, D. (1989) *The Urban Experience*. Oxford: Blackwell.

Hight, J. (2013) 'Narrative archealogy', in Buschauer, R. and Willis, K.S. (eds), *Locative Media: Multidisciplinary Perspectives on Media and Locality*. Bielefeld: Transcript Verlag.

Hill, D. (2012). 'In praise of lost time.' *Domus*. http://www.domusweb.it/en/design/2012/03/05/in-praise-of-lost-time.html (Retrieved: 1 August 2014).

Jeffries, A. (2013) 'Taxi race: can Uber and Hailo deliver a real-time revolution?' *The Verge* http://www.theverge.com/2013/2/7/3964394/taxi-race-can-uber-and-hailo-deliver-a-real-time-revolution (Accessed: 6 August 2014).

Kern, S. (1983) *The Culture of Time and Space, 1880–1918*. Cambridge MA: Harvard University Press.

Knorr-Cetina, K. (2003) 'From Pipes to Scopes: The Flow Architecture of Financial Markets'. *Distinktion*, 7 pp. 7–23.

Lash, S. and Urry, J. (1994) *Economies of Signs and Space*. London: Sage.

Lefebvre, H. (1991) *The Production of Space*. Oxford and Cambridge, MA: Blackwell.

Lefebvre, H. (2004) *Rhythmanalysis: Space, Time and Everyday Life*. Translated by Stuart Elden and Gerald Moore. London, New York: Continuum.

Ling, R. and Yttri, B. (2002) 'Hyper-coordination via mobile phones in Norway', in Katz, J. and Aakhus, M. (eds), *Perpetual Contact: Mobile Communication, Private Talk, Public Performance*. Cambridge: Cambridge University Press, pp. 139–169.

Lynch, K. (1988) *What Time is the Place?* Cambridge, MA: The MIT Press.

Lyster, C. (2013) 'The Future of Mobility: Greening the Airport'. *Places Journal*, July 2013.

Macleod, I. (2013) '89% browse or buy using smartphones and tablets during their commute'. *The Drum*. http://www.thedrum.com/news/2013/12/04/89-browse-or-buy-using-smartphones-and-tablets-during-their-commute (Accessed: 25 August 2014).

Marchetti, C. (1994) *Anthropological Invariants in Travel Behavior, Technological Forecasting and Social Change* http://www.cesaremarchetti.org/archive/electronic/basic_instincts.pdf: International Institute for Applied Systems Analysis, Laxenburg, Austria, pp. 75–88.

Mawoud, J. (2014) 'Dubai, Once a Humble Refueling Stop, Is Crossroad to the Globe'. *The New York Times*. http://www.nytimes.com/2014/06/19/business/international/once-a-humble-refueling-stop-dubai-is-crossroad-to-the-globe.html?_r=2, JUNE 18, 2014.

Mayer-Schonberger, V. (2009) *Delete: The Virtue of Forgetting in the Digital Age*. Princeton, NJ: Princeton University Press.

McLaren, C. (2011) 'The Senseable City: An Interview with Carlo Ratti', in *LabLog*. http://blogs.guggenheim.org/lablog/the-senseable-city-an-interview-with-carlo-ratti/: Guggenheim. 2014.

Mels, T. (ed.) (2004) *Reanimating Places: A Geography of Rhythms*. Aldershot: Ashgate.

Minute Suites. www.minutesuites.com (Accessed: 1 October 2014).

MIT SENSEable City Lab (2006) *Real Time Rome*. http://senseable.mit.edu/realtimerome/ (Accessed: 1 October 2014).

Mitchell, W.J. (1995) *City of Bits*. Cambridge, MA: The MIT Press.

Mobile in the Airtravel Industry Report 2014. (2014) http://www.eyefortravel.com/mobile-and-technology/mobile-airtravel-industry-report-2014: Eye For Travel.

Pascoe, C. (2011) 'Time, technology and leaping seconds', http://googleblog.blogspot.de/2011/09/time-technology-and-leaping-seconds.html (Accessed: 2014).

Pew Internet Project (2014) 'Social Networking Fact Sheet'. [Online]. http://www.pewinternet.org/fact-sheets/social-networking-fact-sheet/.

Rheingold, H. (2003) *Smart Mobs: The Next Social Revolution*. New York: Basic Books.

Robinson, D. and Davies, S. (2014) 'Online sales lift small takeways up the pecking order'. *Financial Times*. http://www.ft.com/cms/s/0/dacdea8c-bb3a-11e3-b2b7-00144feabdc0.html-axzz3NwaVZUzL, 3 April 2014.

Rosen, J. (2010) 'The web means the end of forgetting'. http://www.nytimes.com/2010/07/25/magazine/25privacy-t2.html?pagewanted=all&_r=0, 21 July 2010.

Schofield, H. (2014) 'Paris car ban imposed after pollution hits high'. *BBC*. http://www.bbc.com/news/world-europe-26599010, 18 March 2014.

Sheller, M. & Urry, J. (2000) 'The city and the car'. *Int. J. Urban Reg. Res.*, 24 (4), pp. 737–757.

Sheller, M. and Urry, J. (2006) 'Introduction: Mobile Cities, Urban Mobilities', in Sheller, M. and Urry, J. (eds), *Mobile Technologies of the City*. Oxford: Routledge, pp. 1–17.

Simmel, G. (1971) *On Individuality and Social Forms: Selected Writings [of] Georg Simmel*, ed. Levine, D.N. Chicago and London: University of Chicago Press.

Sullivan, M. (2011) '20 Best U.S. Airports for Tech Travelers'. *Computer World*. http://www.computerworld.com/article/2500425/enterprise-applications/20-best-u-s--airports-for-tech-travelers.html?page=2 (Accessed: 1 October 2014).

TaskRabbit *How Does TaskRabbit Work?* https://www.taskrabbit.com/how-it-works (Accessed: 1 January 2015).

The Economist (2013). *The rise of the sharing economy*. http://www.economist.com/news/leaders/21573104-internet-everything-hire-rise-sharing-economy (Accessed: 1 August 2014).

Thrift, N. and May, J. (2001) *Timespace: Geographies of Temporality*. Routledge Critical Geographies Series. New York: Routledge.

Townsend, A. (2000) 'Life in the real-time city: mobile telephones and urban metabolism'. *Journal of Urban Technology*, 7 (2), pp. 85–104.

Twilio. https://www.twilio.com/elements (Accessed: 1 October 2014).

US Air Force (2014) *Timing*. http://www.gps.gov/applications/timing/ (Accessed: 1 October 2014).

Wasik, B. (2010) *And Then There's This: How Stories Live and Die in Viral Culture*. New York: Penguin.

Whyte, W.H. (1980) *The Social Life of Small Urban Space*. Washington, DC: The Conservation Foundation.

6

Things

6.1 NETWORKED THINGS

Home

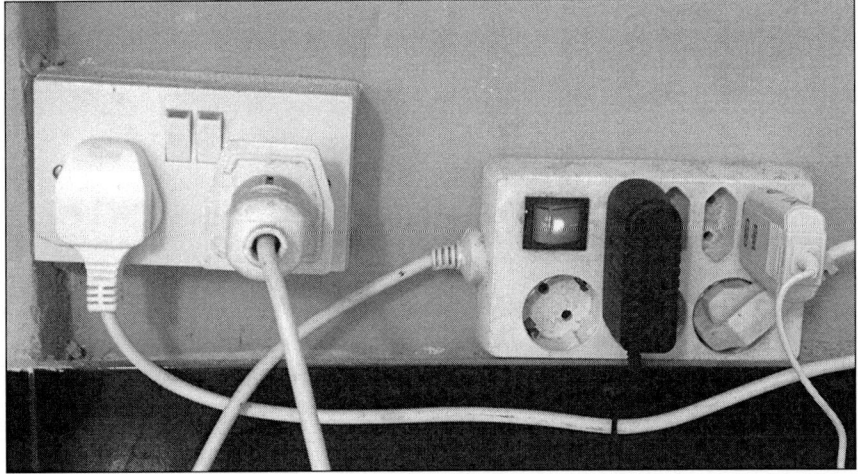

6.1 Nine plug sockets in the kitchen in my home.

There is a common saying: 'home is where the heart is'. My home is also where 34 electric plug sockets, 1 sink, 1 oven, 1 fridge, 15 electric lights, 1 boiler, 1 washing machine, 1 tumble dryer, a toaster, a kettle, 2 TVs and a Wi-Fi router are (Figure 6.1). If I were to remove the latter two on that list from our home, then my husband and especially my two sons would probably clamour for their immediate return, since for them the 'heart' would be gone. In fact I've noticed that each room in the house tends to be increasingly orientated, not around the 'hearth', but around these appliances; the TV is the focal point of the living room, the fridge, dishwasher and oven take centre stage in the kitchen, alarm clocks and plugs feature in the bedroom, the electric shower is key in the bathroom and not forgetting the central presence of the wireless router in the hall. The array of electrical devices spread throughout the house, all require controls, switches, plugs and of course power. More generally, the radiators and electric lights that operate through the house, are connected through a basic system of hard-wired pipes and electrical wires that weave through the hidden spaces between the walls and under the floor. Every day I am made aware of the materiality of theses devices; the plastic surfaces that gather dust, the plugs that are not in the places where I actually want them, the vacuum cleaner, which when switched on, drowns out the sound from the television. None of them actively connect to each other or give me information about their state or consumption. They are opaque boxes of functionality that require maintenance, cleaning, storage and power; they are messy, fragile and can become obsolete. They also draw on different natural resources; the oven needs gas, the washing machine needs water and the fridge and TV need electricity in order to function. I have no idea how much electricity my fridge uses, or the tumble dryer (which I expect is the most greedy in terms of energy use) or the TV. But their use of resources is manifested in the monthly bills; it has a value that is materially quantified at a series of meters housed in a cupboard under the stairs. The reading of these meters requires a human; this is either myself, or someone from the relevant company, who visits the house to note down a series of numbers. Once recorded, my consumption is calculated and appears on my bill; the flows of water, gas and electricity into and out of the house are translated into tangible units that are given monetary value. The house is also pretty analogue in terms of access and security; I use a key to get in and out, and if I lose it then the simple fact is that my house will not allow me to enter. When I am safely at home, a doorbell signals a stranger at the door, and I have the choice as to whether to let them in. The only device in the house that lets me in and with which I can connect remotely is the Wi-Fi router. However the access is highly coded; I need a fourteen-digit, alphanumeric password, and have to enter through a local Internet protocol (IP) address in my Internet browser. The router is also the only device that not only brings flows into the house but also enables me to transfer data back out through the fibre optic cable. In terms of equivalence, the washing machine does connect to the wider world when it discharges dirty water into the sewage system, but this is a one-directional flow; there is no feedback or connection with the water that enters the house. With the router, my home acts as a point of exchange and connection with the wider world; I upload files, send out emails and call via Skype. For many the 'heart' of the home is the about a sense of identity created by containing and controlling what goes on inside a domestic setting; close family relations, and ownership of 'things' and spaces that instil a sense of personal identification with the setting. But increasingly the 'heart' of a home is a beating pulse of intangible digital data, and this is what defines a sense of who we are.

Things, spaces and identity

A home can be a material shell of bricks and mortar, as well as a metaphorical heart. We make sense of the material world based not just on the material properties but on how we identify with it. Our sense of self is about constructing a sense of what we are not, and being aware of what we identify with; the things that frame our everyday world, as well as the memories of the past and hopes for the future that we project onto them. A room is full of material things or 'stuff'; chairs, tables, pictures and lights, but they are also shaped to define a sense of self. Increasingly the 'stuff' that makes up our world is not material in the tangible sense of physical objects; it is comprised of information or data. In fact if we look around information is everywhere. It is in the air around us in the form of Wi-Fi, radio and mobile signals; it is in the cables beneath our feet carrying Internet data; it is in the telephones and laptops that we carry with us. This manifests itself in other ways in which we construct and maintain our sense of identity; a Facebook page, a particular mobile phone and digital photo archives. These are not 'things' in the sense that they can be physically touched or owned, but they are 'thing-like' in that they are a core part of how we construct our identity with the material world; part of our agency. As the anthropologist Miller points out; 'people still strive to construct relationships to people and things. These relationships include material and social routines and patterns which give order ... an order which, as it becomes familiar and repetitive, may also be a comfort to them' (2010, p. 296). We construct relationships around, with and between digital 'things'. Yet many of the ways we talk about technology suggest the 'digital derives its power from its nature as a mere collection of 0s and 1s wholly independent from the particular media on which it resides (hard drive, network wires, optical disk, etc.) and the particular signal carrier which encode bits (magnetic polarities, voltages, or pulses of light)' (Blanchette, 2011). In this characterisation of technology, it is immaterial or lacking 'thingness'. But increasingly we experience technology in many material ways; and more and more it is part of how we construct an identity for ourselves, and also for how we connect to and with the world around us.

Senses and sensors

The forming of identity is fundamentally a situated process; perceptions of self, identity, and memory are inextricably linked with our sense of belonging in a spatial setting. As such identity is in part a quality that exists outside of a specific time, but is a result of experience. Lynch and Gibson drew on the field of environmental psychology to formulate an idea of identity as being formed through perceptual experience (Gibson, 1979; Lynch, 1967). In the 1960s academics from the field of psychology introduced the idea that perception frames our identity and understanding of the environment around us. Lynch highlights how making sense of the world is a very visual experience (1967) whereas Gibson introduced the idea of affordances (1979). This concept outlined how the properties of the world around us enable different actions, such that 'an action possibility available in the environment to an individual, independent of the individual's ability to perceive this possibility' (McGrenere and Ho, 2000). Norman (1998) reinterpreted Gibson's concept of affordances to include 'the perceived and actual properties of the thing, primarily those fundamental properties that determine just how the thing could possibly be used. [...] Affordances provide strong clues to the operations of things. Plates are for pushing. Knobs are for turning. Slots are for inserting things into. Balls are for throwing or bouncing' (p. 9). This outlines an approach to the world around us being experienced through the senses; sight, hearing, smell, touch and taste, that enable us to perceive and experience the world beyond the body. These sensory inputs are somehow converted into perceptions of desks and computers, flowers and buildings, cars and planes; into sights, sounds, smells, taste and touch experiences. Much of this thinking is based on the core foundation that the human body is somehow physically and perceptually separate from the external world and its objects. This also precludes the concept that the world around us has materiality and form; it can be perceived because of its 'thingness'.

Recent approaches in both theory and technological development have prompted a challenge to this understanding of the externality of materiality and form. Approaches such as Actor Network Theory (Callon, 1986; Latour, 1987; Law, 2002) have introduced the framework that human actors and objects, as well as the spatial world are linked through network relationships. ANT argues that non-humans have the capacity to be actors or participants in networks and systems. In terms of materiality, developments in the links between digital processes and product design for example have led theorists and practitioners such as Dunne to argue that 'the electronic object is on the threshold of materiality. It encourages a focus on experiences, not on tangible objects' (1999, p. 11). More recent is the emergence of the field of Internet of Things (IoT) – a world of sensors, flows, things talking to each other, and everything being tracked and visible. This scales up the urban through the concept of the 'sentient city' (Shepard, 2011) and sentient cities (Crang and Graham, 2007) that Shepard describes as the 'dataclouds of 21st century urban space' that shape our experience of the city (2011). According to de Waal 'all over the city, 'intelligent' applications have started sensing what is happening around them and reacting to it – be it smart traffic lights or CCTV camera's whose images are computer analyzed for suspicious behavior. Add to this the increase of tracking devices such as cell phones that most urbanites carry, and as a result the city has become 'sentient'' (2011). One of the consequences of the emergence of a 'sentient city' is the changing nature, materiality and value of 'data'. Increasingly the network of people and objects is not just an immaterial, flowing process, but a quantifiable set of information that is not only created, but also captured, stored and then processed. This is 'Big Data', which results out of the increased ubiquity of sensors and the proliferation in digital storage capacity (Mayer-Schonberger and Cukier, 2013). However what Big Data suggests is that it is no longer the human that senses the world, as an array of ubiquitous and connected sensors within the world also affect how we experience environments and things. If we reflect this thinking back to look at the home environment however, problematic situations and relationships could emerge when, for example, my fridge knows that I've run out of milk and orders it for me even though I'm going to be away for a few days, or where there is no longer a switch to change the light levels in a room since it is all controlled through sensors.

Software and hardware

Some of the core arguments around the emergence of digital networks are based around claims that such networks dematerialise the 'real' world. The 'virtual' world, it is argued, is transparent, placeless and lacking identity. The introduction of a series of perspectives such as 'Hertzian space' (Dunne, 1999) and the notion of 'tangible bits' (Ishii and Ullmer, 1997) sought to give more definition to the world of objects that exist between the virtual and the physical. These approaches highlighted the material qualities of digital 'bits' and sought to understand their material and tangible properties. According to Dunne 'the electronic object occupies a strange place in the world of material culture, closer to washing powder and cough mixture than furniture and architecture' (1999, p. 16). More recently the question of the materiality of the digital world has been further brought into focus through the emergence of technologies characterised as IoT or Big Data, which not only sees objects as being digital, but also connected. According to Shepard the emergence of connected things, space and people 'requires thinking about space in non-visual ways, where formal geometry and material articulation become less relevant than the topologies of networked information systems and their intersection with the socio-spatial practices of daily life' (2010). A number of authors have highlighted that digitally mediated objects and spaces are not just perceived as 'hardware' but are also defined increasingly by the software. Kitchin and Dodge (2011) introduce the term code/space to outline the social and spatial effects of software embedded within the material world, and also to explore ideas of agency within such systems where the distinction between human operated and digitally automated is increasingly ambiguous. This raises broader questions of human agency within such code/spaces, and also how this translates into a sense of human identity. When the material boundary between people and objects becomes blurred, so the psychological framework that we use to make sense of the world becomes challenged. For example, if we return to the case of the home environment introduced at the beginning of the chapter; then we can see scenarios where a home could start to function autonomously of its occupants; sensing, recording and storing data. This causes the home to take on a different relationship with its owner. If part of how we make sense of ourselves is through how we choose to construct and maintain relationships with the material and physical spaces and objects, then the connected and 'sentient' home reframes some of these basic ideas.

6.2   DIGITAL AND MATERIAL WORLDS

Identity and our relationship with the material world

Rapoport (1981) has documented how the form and location of a house represented social status and group membership. He argues that people construct meaning and a sense of identity through the material world in a process where 'material objects first arouse a feeling that provides a background for more specific images, which are then fitted to the material ... so that the physical environment – clothes, furnishings, buildings, gardens, streets, neighbourhoods and so on, is used in the presentation of self' (1982, pp. 14–15). Rapoport highlights the importance of meaning making in our relation with material objects. In doing so, he does not focus on the importance of the physical qualities or properties of these objects, but rather on how they come into being through our relationship with them.

The challenge of the digital world that has emerged over the last forty to fifty years, is characterised by Negroponte as a shift 'from atom to bits' (1995). The change in the last twenty years towards 'ubiquitous computing' has realised Mark Weiser's dream of invisibility and immateriality. According to Weiser (1991) 'the most profound technologies are those that disappear. They weave themselves into the fabric of everyday life until they are indistinguishable from it'. Weiser uses the example of electric motors to illustrate how this process has happened historically where 'At the turn of the century, a typical workshop or factory contained a single engine that drove dozens or hundreds of different machines through a system of shafts and pulleys. Cheap, small, and efficient electric motors made it possible first to give each machine or tool its own source of motive force, then to put many motors into a single machine' (Weiser, 1991). We have now reached a fairly recognisable version of Weiser's vision of ubiquitous computing embedded invisibly into the world around us. Computer chips and computing capacity are woven into the material world of the built environment and the objects within it, invisibly. The question raised by this changing materiality of data is how we use them to construct ideas of self and identity.

Data clouds

Every day we make multiple and various interactions with the data cloud. The result of software-connected data is what Hill describes as 'urban data clouds' made up of 'a new kind of data, collective and individual, aggregated and discrete, open and closed, constantly logging detailed patterns of behavior' (2008). Our interactions with this 'data cloud' are sometimes based on decisions; that is we choose or have agency in how we present, save and share the data. Yet many are not; data gathering is automated and results in further, often unknown, triggers for other machine-to-machine interactions that share and distribute the data. In fact every day we create approximately 5,000 megabytes, in an exponential curve where 'every day of your life, more data is being uploaded than created throughout all recorded history until a couple of years ago' (Scoble and Israel, 2014, p. 5). This data used to be held in silos, where as individuals we still had some thread of ownership or control over how it was accessed and stored. So for example, in the seventies our bankcard details were maintained by a particular branch of the bank, and were not passed on to third parties. In the context of the rise of 'Big Data', Kitchin and Dodge (2011) distinguish code/space from 'coded space'; which is where software augments the experience of space but remains peripheral. Code/space, they argue, articulates a new set of assemblages where software is increasingly an active agent in capturing, processing, storing and sharing our interactions with the world around us.

The materiality of digital objects

Despite the promise of invisibility and immateriality digital objects operate on the boundary between material and digital. In fact although our 'sense' of the presence of digital objects is that they lack material presence, all digital devices and objects still require materials: computer chips, optic fibres, copper wires, cables, cell phone masts, and, fundamentally, the silicon. These give materiality to immaterial data. Dourish and Mazmanian argue that data can be brought back into the realms of experience by a focus not on materials, but on materiality (2011). By reclaiming the range of material properties of digital objects and data 'questions of durability, fragility, visibility, malleability, deformability, density and heft that contribute to the sociocultural considerations' (Dourish and Mazmanian, 2011, p. 5). Similarly, whilst digital data gets smaller in size despite growing in capacity, energy use is conversely on the rise. Almost any digital device or process requires an energy supply, generally in the form of electricity. Anyone who has searched for a plug for a laptop, phone or tablet will recognise the contrast between the mobile and wireless capabilities of the device and immoveable location of a power supply. Graham argues that energy use is an example of 'the material vulnerability that underlies the so-called 'virtual world'' (Farias, 2011, p. 199). It is through these sorts of material processes that intangible and invisible data and digital objects become exposed as material; through energy use, when they decay and when they require maintenance. By engaging more actively in these characteristics of electronic objects the point at which information can become material in the context of our everyday life is more realisable.

Aims

This chapter aims to explore how the immateriality of data affects the material world of the built environment, in the context of the increasing connectivity between objects, spaces, things and data. It starts out be exploring the concept of identity and how the material world shapes our sense of self, and then reflects on the sentient nature of the city. The first section deals with ideas of materiality and looks at how the senses are often used as a framework to interpret experiences, and contrasts this with a digitally constructed and coded system of materiality that introduces a number of ambiguities. It looks at how ideas of wireless connections and data as material are actually still intimately interwoven with the material world, and that issues of storage, waste and memory construct material frameworks around immaterial flows. The chapter uses the lens of the home, a place where we most closely construct a sense of our identity in the world, and tests out the changing boundaries between people, places and data through history to understand the current changes in a broader perspective. The social context of material things and electronic objects is discussed, as well as issues of agency. I then focus in on a range of changing conditions and explore these in more detail through an investigation of a case study which looks at the 'smart home'. The way that technologies characterised as IoT and smart sensors affect the way that the domestic environment is experienced is also discussed. Finally the chapter explores the implications of these findings for the field of architecture, and in particular focuses on the challenge of the relationship between code and space, as well as the opportunities for different ways of thinking about materiality and form within the built world.

6.3 HISTORICAL CONTEXT

Technological objects: from crafted tools to sentient things

The Great Exhibition of 1851, held at Crystal Palace in Hyde Park, London, UK attracted over one third of the nations population. It showcased the technology and materials at the turn of the industrial age; an envelope machine, kitchen appliances, a voting machine, steel-making displays, a reaping machine, and even a precursor to the modern fax machine. This was alongside material wonders; the world's largest diamond and examples of fine materials such as flax, silk, lace and cotton from all over the world. These were all housed within a material wonder itself; the lightweight glass and steel framework of the Paxton designed Crystal Palace, measuring 564 metres long by 138 metres wide. The Great Exhibition heralded the twentieth century's shift into industrialisation, and the corresponding influence on society of mass design processes, machine production and the potentials of new materials such as concrete, steel and glass. This was the case in architecture and also the fields of product and industrial design that emerged in the first decades of twentieth century.

Industrialisation provided the foundations for the modernist movement, which emerged out of a fascination with technological progress and the potential of new materiality of production. According to Le Corbusier, one of the key exponents of modernism; 'mechanisation has called forth a new spirit ...' (1989, p. 147), and he contrasted the streamlined, engineered shapes of ocean liners with what he considered to be the unnecessary materiality of thick, solid walls. The home became one of the domains seen as having the possibility of deriving most benefit from engineered forms, materials and processes. Le Corbusier declared that 'a house is a machine for living in. Baths, sun, hot water, cold water, controlled temperature, food conservation, hygiene, beauty through proportion' (1989, p. 151). The modernist ideal of home became a site for mechanisation and engineered efficiency, inspired by the products and materials of the industrialised aged. It also became about control of the home environment; of water, heat, light as well human activities (prescribed by functions), in fact 'the emphasis was on literally overpowering the natural environment with mechanical technology (mur neutralisant)'(Mackenzie, 2011).

In the 1960s, the introduction of computers led to a new approach towards creating buildings that responded or were controlled, but this time the control was reimagined as being executed by their users. Negroponte proposed a 'responsive architecture' which integrated computing power into built spaces and structures, with the idea that this could improve their 'performance'. Negroponte and others foresaw an environment: that 'would not only be able to monitor and regulate environmental conditions but also to mediate the activity patterns through the allocation of functional spaces' (Wellesley-Miller, 1976). Inherent within this approach was the idea that the building starts to 'know' the inhabitant and is able to respond to their behaviour. According to Grunkanz (2012) 'the common introduction to responsive architecture is usually made by using the example of the thermostat. It is a basic example of a cybernetic feedback loop placed in a building environment in which the actual output is affected in response to an input'. Buildings and objects were no longer simply adjusting to the natural environment but to the behaviour of the people and things within it, through the introduction of sensors and feedback loops. In the last twenty years, the reduction in size of computers has led to objects and materials where computing power is not just linked to, but embedded within them. In the late nineties, Ishii and Ulmer introduced the concept of 'tangible bits' (1997); 'dynamic physical (and computational) material that can conform, transform, and inform' (Ishii et al., 2012), which was a precursor of the emergence of the Internet of Things. This moved into an era where the distribution of technology within everyday objects and environments was ubiquitous; the material of technology has moved from the engineered machinery of the early twentieth century to a sensor-embedded and internet connected system of input and feedback loops.

Social objects: from tools to interactions

Industrialisation brought about a change in not just the way that technology could be used in the material production of objects and environments, but also in its social production and use. This saw a transformation in the way tools and technology were socially constructed; with a shift from man-made tools to automated and complex machines, and a consequent shift in the role people played in their relationship with technology. However one of the problems with the modernist utopian vision of technological progress that has been discussed, is that lack of importance placed on human-machine relations, and the lack of appreciation of the socio-cultural context (Feenberg, 1999). In the seventies researchers from the field of Psychology brought in a way of understanding the human object relationship such as the concept of affordances, introduced by Gibson. Gibson stated that 'an affordance is not bestowed upon an object by a need of an observer and his act of perceiving it. The object offers what it does because it is what it is' (1979, p. 138). The concept of affordances stressed 'relevant human-scaled objects, attributes and events and the patterns of energy that provide effective perceptual information' (Gaver, 1991, p. 79).

In the eighties, social scientists started to reassess the way that human relationships with technology were constructed and understood. Suchman, an anthropologist employed by Xerox Parc (where Weiser's concept of ubiquitous computing emerged) documented the problems that arise when technology lacks an understanding of the social context. She argued that there needed to be room in human-machine relations for the person to construct their own frameworks of use. Suchman investigated the use of a large and complex photocopier, which despite being technically very advanced, turned out to be almost unusable by those it was designed for. She pointed out that that people's behaviour is contextualised, and argued that it was important to understand the situation in which they are acting in order to determine what people will do (Suchman, 1987). A complimentary approach emerged in the eighties in the field of 'interaction design', which addressed the context of designing for 'all the interactions that are enabled by digital technology, whether by computers, chips embedded in products or the environment, services or the internet' (Moggridge, 2007, p. 660). Critically, interaction design saw these interactions not just as socially constructed and everyday; they also addressed the challenge of how such interactions might be emotive and sensory. A number of approaches, such as 'embodied interaction' (Dourish, 2001), 'affective computing' (Picard, 1997), and 'tangible computing' (Ishii and Ullmer, 1997), have increasingly focused on the emotive and sensory experience of computing and opened up the challenge of understanding and working with the social context of use. These approaches see the potential of constructing meaningful experiences with computing as being grounded in embodied and sensory human experience and in the social context of everyday use.

6.4   CONNECTED OBJECTS, PEOPLE AND SPACES

The socio-cultural context of sentient objects and cities

One of the key challenges of the invisibility and immateriality of ubiquitous data is that it privileges invisible interaction with data, through interfaces such as touchscreens, swipe cards, voice and even face recognition. They leave little room for people to construct ways of interacting with technologies that adjust to the socio-cultural context. At the scale of the city, Shepard (2010) describes how 'sentient cities' need to be understood not as abstract data flows, but as 'assemblages of code, people, and space' that 'are brought into being through specific techno-social performances or enactments within the course of daily life'. In his discussion of the concept of Hertzian Space, Dunne (1999) argues that more focus on socio-culturally constructed meaning making is required in order for people to construct meaningful relationship with electronic objects. These narratives occupy the space between rationality and reality, and challenge the apparent seamless functionality of the digital object (Dunne, 1999, p. 56). It is through such narratives that identity construction can become part of the relationship we construct with electronic objects and data. One way in which this can be achieved is to think about how sentient cities or objects might be what Rose characterises as 'enchanted objects' (2014), that is objects that can transform, conjure, and invoke. As more of our objects and environments become actuated, connected, and data-enabled, these enchanted objects are developing the capacity to contain their own stories or narratives. An object can remember its history, can understand how it is used, and can talk to other objects around it to understand its environment. As the capabilities of sentient objects and environments evolve, objects and spaces no longer become inert backdrops to our experiences, but active participants in our world that can share stories about themselves and us. The question is what sort of relationships do we form when these objects and spaces take on lives with emotional capacity when this is combined with computing power and connectivity to actually act on their responses? Will out houses talk back if we make a mess, refuse us entry if we have a major haircut or try to comfort us if they sense we are feeling upset?

Code and agency

One way in which agency is performed in our interactions with digital objects and spaces is through software. Kinsley highlights how 'software programmes thus have significant agency in the various ways in which we collectively communicate and remember' (2014, p. 2). Kitchin articulates a typology of code/space; 'automated' where data is collected by software-driven systems, 'directed', where data is collected by systems controlled by a skilled operator; and 'volunteered', where data is submitted voluntarily, perhaps in exchange for value provided by a service based on those data – such as social media, or store loyalty cards. Whether we volunteer our data, or whether it is automatically gathered without our knowledge, our lack of agency in how it is processed leaves us vulnerable to systems where our relationship with our data is contested. For instance Graham (2005) introduces the practice of what he terms software sorting where 'less lucrative users of streets, mobility systems, services, electronic communications grids, and places can be electronically (and/or physically) pushed away and marginalised, either absolutely or relatively, through software-sorting and machinations of code' (2005, p. 566). Here the agency is no longer in the person who created the data, but is now performed, at least in part 'through the continuous agency of vast realms of computer software' (Graham, 2005, p. 562). There is a shifting of roles in this process; for software sorting requires the person to be described in terms of software, or more precisely metadata. For example to a number plate recognition system, I am are not a forty-two year old mother, who has just finished work; I am a Honda Jazz, registration WJ04 FXC and built in 2004. This metadata does not contain personal or content-specific details, but rather transactional information about the user, the device and activities taking place. Until we can find ways to resist, reclaim or make agency in how we present ourselves, and the objects in our everyday world, in terms of software and code then we are at risk of losing claim to some of the agency and sense of identity we have as flesh and bones humans.

Data histories and memories

One of the ways that material objects take on both a material quality and also contribute to a meaningful relationship with the particular social-cultural context is through storage. Although data is often characterised as ephemeral, the rise in sentient technologies means that it also has a lifespan and a history. One of the more individually focussed set of IoT technologies is those that support the construction of what has been termed the quantified self; from fitness apps that monitor daily exercise, to the measurement of sleep patterns and more recently the quantification of emotion. Almost everything we do generates data, but it is increasingly recognised that data has lives or 'biographies'. In fact one of the promises of the rise of sentient technologies is that things, objects and issues can record their own 'biographies'. Biographies of things can give meaning to what might otherwise remain obscure (Kopytoff, p. 66) since they can acknowledge the diverse encounters between people and things as they move through time and space. This sees data as contributing to a 'social life of things' (Appadurai, 1997) that leaves traces of 'the forms, uses, and trajectories of things-in-motion' (Dwyer and Jackson, 2003, p. 270) that can reveal their human and social context. Bratton and Jeremijenko state that the process of revealing data biographies may not be sufficient, and that this process needs to include the possibility of opening up the data to questions about its creation, use and possible future trajectories. They argue that to enable a sense of agency we need to ask the following: 'do these projects change who is asking the questions? Are these designers now asking the question of how this pollutant is made, who made it, where is it coming from, where is it going, what do we do about it, or not? ... Who collected [the data] and under what conditions' (Bratton and Jeremijenko, 2008, p. 11).

6.5   CASE STUDY: SMART HOMES

In this case study I use the setting of the home to explore the impact of sentient objects and things the way in which we construct a sense of identity with our spatial and material world. Miller describes houses as 'the elephants of stuff' with 'hard strong material presence' but he also argues that the 'values and meanings objectified in housing are themselves subject to change' (2010, p. 81). In this context the promise of the smart home, full of sensors and connected both within itself and beyond, offers up new potentialities for how people live and the consequent values and meanings objectified within the setting of the home.

The heart of the sensory home

Smart homes incorporate a range of sensors embedded within everyday objects and services that are linked through either a Wi-Fi or a cabled connection. These different systems tend to be managed and maintained centrally through some form of control system, which can be accessed through an interface linked to a mobile phone or tablet 'app' (Figure 6.2). According to a 2014 report one in eight European homes could be 'smart' by 2019 (Berg Inisght, 2014). The report groups smart homes into six primary categories: 'energy management and climate control systems; security and access control systems; lighting, window and appliance control systems; home appliances; audio-visual and entertainment systems; and healthcare and assisted living systems' (Berg Inisght, 2014). Irrespective of the particular technological configuration of a smart home, its purpose – according to

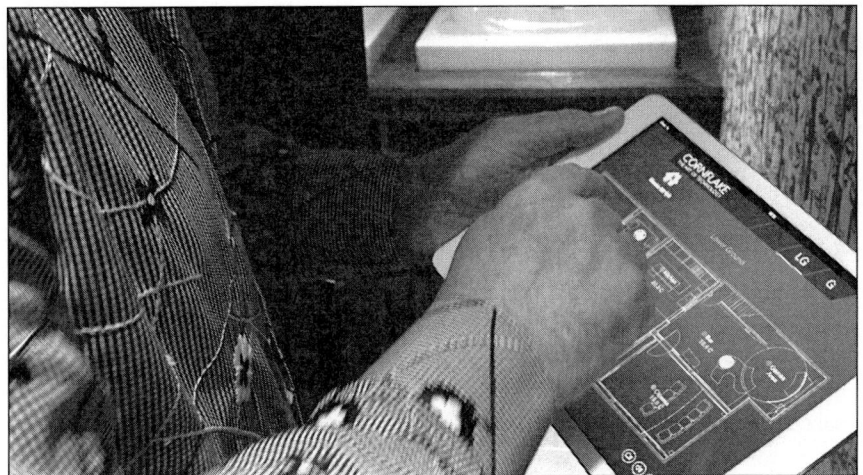

6.2 iPad interface for 'Smart App-artment', London, 2014.

technology developers is 'to improve the living experience' in some way (Gracanin et al., 2011). This language of the smart home focuses on 'systems', 'efficiency', and 'control', all words that seem at odds with the material and socially constructed ideas of home. In a recent study Pink has offered another reading of the smart home which she terms the 'sensory home'. Pink's team undertook ethnographic studies of how occupants made sense of energy use and technology within their home and found that 'the making and experience of the home as a multisensory environment was likewise integral to how self-identities are constituted through everyday life practices' (Pink and Mackley, 2012). In fact further studies focusing on the owners of smart home technology have noted the multiple ways in which they actively construct and maintain sensory relationships with technologies such as remotely controlled vacuum cleaners, thermostats and entry systems. For example Sung et al. undertook a study of the owners experience of Roomba; a vacuum cleaner robot. The team found that 'The strength of the relationship that our participants felt with their Roombas not only encouraged them to promote Roomba to others, but also motivated them to modify their living environment to accommodate the floor vacuum' (2007, p. 154). Not only did people feel a strong emotional connection with their Roomba, but also they actively modified their home to accommodate it. Sung recounts how 'twenty-seven of the 30 households we spoke with had made changes to their houses to accommodate Roomba' (2007, p. 154). The key factor was that, although the vacuum cleaner operated as an autonomous agent, the presence of Roomba within the intimate setting of the owners domestic environment meant that they constructed an empathetic relationship with it. This led them to adapting and even quite radically changing their routines in a process that Sung et al. characterise as 'Roombarisation', where 'the presence of intimacy opens up new possibilities for how people will incorporate this technology into their home routines' (2007, p. 157). This included some owners going out at certain times so that Roomba could work, or others removing awkward objects such as floor rugs to 'help' the Roomba do its job. According to one user 'when we know the Roomba is going to be cleaning the next day, we don't want that stuff to get in the way so we tend to put things away more.' Another commented 'I can't imagine not having him any longer' (Sung et al., 2007, p. 150, 153), and in this way they constructed a mutual social-spatial narrative within which they framed their domestic lives with Roomba.

The networked house and agency

In 2013, UK technology company Cornflake, created a 'Smart App-artment' in the centre of London (Cornflake). The two-floor apartment features a spatially standard UK apartment, with a kitchen, a living room and bedrooms. However it is fitted with an astonishing array of automation; security, lighting, energy entertainment and even cleaning are delivered through automated systems operated by the house owner through a tablet device. Much of the way the house is experienced is through screens. For instance according to a press article 'the Cornflake bar room also has a 'flat frog' screen, shown in the foreground, which is a very high-end multitouch table that four or five people, such as a whole family, can use at the same time. It can recognise and distinguish each separate pair of hands. Images from the table can be beamed to the projector, shown in the background' (Steiner, 2013). Security is also a major aspect of 'App-artment' design such that the space 'includes a secret wall and floor that rotates to reveal a hidden two meter television screen and a virtual security guard complete with Alsatian 'patrolling' various rooms by projector' (Steiner, 2013). Domestic routines are also managed through automation so that when the occupant is away 'an LED Growlight, pictured, hidden beneath wall units keeps plants and herbs alive while a watering system can be operated remotely while on holiday' (Steiner, 2013). The extent of the computing power required to maintain the networked connectivity is made evident in the fact that a separate, highly serviced room is required to accommodate the servers that run the control systems. All this gadgetry comes at a significant cost. For example the hot drinks maker operated remotely by an iPad costs $15,000 or £10,000, whilst the array of specialist technology such as magnetic wallpaper, a cinema and other high-tech gadgetry run up a total cost of $1.1 million or £750,000. That's just for the technology. In fact the current array of smart home appliances and systems all seem to come at a significant cost. A Roomba currently costs around $500 (irobot.co.uk), and an the LG Smart ThinQ washing machine that allows you to 'monitor your laundry remotely', download new wash cycles and self-diagnose any problems costs around $1600 (LG, 2015). This high cost of almost every feature of the 'smart home' excludes a vast percentage of homeowners, and particularly those on lower incomes who could potentially benefit from the 'efficient' energy saving capabilities of such technologies. But on a more fundamental level, the high level of automation opens up questions of digital exclusion, and lack of agency in terms of who is 'in control' in the domestic setting. The language and infrastructure of this model of a 'smart home' moves away from the intimacy and routine of a domestic setting and shifts it to the highly serviced and automated model of a space for consumption more in line with a shopping or entertainment venue. One counter to this approach is a more contingent use of smart home technology. Tom Coates created what he has termed the 'Twitter Home' in 2012 using; a series of Belkin Wimo Wi-Fi connected switches that 'allow him to control the lights in his living room, office, and bedroom from his iPhone; a WeMo motion sensor can tell if anyone enters the room; and a Twine device tracks the temperature and the ficus tree's moisture level' (Coates). Coates uses a simple, free Web tool called IFTTT ('If This Then That'), which allows users to set automated online actions in response to defined triggers; such as sending a tweet when a specific person uploads a new Instagram photo. According to Coates, 'you can use IFTTT to say, "When it's sunset, turn all the lights on" – and it'll work that out from where you are in the world … or Start turning the lights off when I ought to go to bed' (Turk, 2013). This set up allows for some forms of communication through the Twitter feed which suggest at emotive responses, such as the following '@houseofcoates – There's a nice lady inside me but she doesn't have a Twitter account so I don't know how to talk to her?' (Turk, 2013)

According to Coates, he still has agency in how the house interacts in the social media platform and points out that 'it's almost like the house has become a sort of pet I look after, and it expresses that being-looked-after-ness back to me … It's like a Tamagotchi or something' (Metz, 2013). In contrast to the Smart App-artment, the Twitter house gives the owner agency quite literally through a set of contingent actions operationalised through the IFTTT platform, and also the intimacy of a Twitter conversation that gives the house a voice within a social media setting.

The materiality of data: dirt, energy use and mess

With such an array of cleaning and optimisation devices, the one condition it is hard to imagine the smart home having to deal with is mess or dirt. According to a recent report by environmental services company Veolia Environmental named 'Imagine 2050' smart, energy-saving homes of the future will 'not need any bins' and will have 'rubbish-sorting robots and a self-cleaning bathroom' (Veolia Environment, 2013). A further report suggests that 'Nanoscopic robots will sort waste in the kitchen and then quickly eat away the rubbish once it has been separated into materials' (Veolia Environment, 2013). Yet interestingly people's experience of the Roomba involved cleaning, not of their homes, but in the maintenance of the robot. According to Sung et al., 'Participants described how brushes, bins and motors needed cleaning to remove the fine dust that might corrupt the sensors and affect Roomba's function. The majority of our participants performed this (approximately 15 minute) task most times they used the robot. This task was the only one that our participants complained about having to do, but unilaterally they preferred this task to that of manual cleaning' (Sung et al., 2007, p. 151). The fact this degree of cleaning is required for a robot that aims to make the cleaning task highly efficient highlights how the mess and dirt of the material world cannot be banished from 'smart' environments, and are, in fact, part of a material framing of our relationship with technology.

6.6   SUMMARY

Our material world is not entirely dissolving into a seething cloud of data, but increasingly built space is becoming embedded with and controlled through network systems. The embedding of highly connected and sensor-based environments often characterised as 'smart', has implications for our 'agency' in the spaces we inhabit. In this chapter I discuss the home as one of the most personal and emotive spaces, and the implications for how a 'smart home' might affect our relationships with the 'things' or material objects or qualities of the domestic environment. A number of authors have characterised these changes as the rise of 'sentient' spaces; picking up on the increasing role of sensors that create, record, process, respond and store data about our everyday spaces. But this approach also raises questions about the degree to which a physical space may be 'sensed' or experienced through the senses, especially when responses with and to the space may no longer be determined by the material and tangible qualities of the space. Interestingly one of the ways that the materiality of such immaterial spaces becomes tangible was shown to be in its maintenance and cleaning; where dirt and mess start to be an important point of encounter with digital processes. In the example at the beginning of the chapter I described the current lack of connectedness of my own home. This will undoubtedly change over the next decade, and the heart of my future home will almost certainly be a hybrid, and probably highly connected merging of the digital and material. With this comes new emotive, everyday relationships with 'things', 'spaces' and 'bits'.

REFERENCES

Appadurai, A. (ed.) (1986) *The Social Life of Things: Commodities in Cultural Perspective.* Cambridge: Cambridge University Press.

Berg Inisght (2014) *Smart Homes and Home Automation.* http://www.berginsight.com/News.aspx.

Blanchette, J.-F. (2011) 'A material history of bits'. *Journal of the American Society for Information Science and Technology,* 62 (6), pp. 1042–1057.

Bratton, B.H. and Jeremijenko, N. (2008) 'Suspicious Images, Latent Interfaces', The Architectural League of New York. http://www.situatedtechnologies.net/?q=node/88.

Callon, M. (1986) 'Some Elements of a Sociology of Translation: Domestication of the Scallops and the Fishermen of Saint Brieuc Bay', in Law, J. (ed.), *Power, Action and Belief: A New Sociology of Knowledge?* London: Routledge and Kegan Paul, pp. 196–233.

Coates, T. @ *House of Coates Twitter account.* https://twitter.com/houseofcoates (Accessed: 1 October 2014).

Cornflake *Smart App-artment.* http://cornflake.co.uk/ (Accessed: 1 January 2015).

Crang, M. and Graham, S. (2007) 'Sentient cities: ambient intelligence and the politics of urban space'. *Information, Communication Society,* 10 (6), pp. 789–781.

Dourish, P. (2001) *Where the Action Is: The Foundations of Embodied Interaction.* Cambridge, MA, and London: The MIT Press.

Dourish, P. and Mazmanian, M. (2011) 'Media as material: Information representations as material foundations for organizational practice'. *Third International Symposium on Process Organization Studies*. Corfu, Greece: 16–18 June 2011.

Dunne, A. (1999) *Hertzian Tales: Electronic Products, Aesthetic Experience and Critical Design*. London: RCA CRD Research Publications.

Dwyer, C. and Jackson, P. (2003) 'Commodifying difference: selling EASTern fashion'. *Environment and Planning D: Society and Space*, 21, pp. 269–291.

Farias, I. (2011) 'An interview with Stephen Graham', in Farias, I. and Bender, T. (eds), *Urban Assemblages: How Actor-Network Theory Changes Urban Studies*. Abingdon: Routledge, pp. 197–206.

Feenberg, A. (1999) *Questioning Technology*. London: Routledge.

Gaver, W.W. (1991) 'Technology affordances'. *Proceedings of the SIGCHI Conference on Human Factors in Computing Systems*. New Orleans, Louisiana, USA: ACM, pp. 79–84.

Gibson, J. (1979) *The Ecological Approach to Visual Perception*. Boston, MA: Houghton Mifflin.

Gracanin, D., McCrickard, D.S., Billingsley, A., Cooper, R., Gatling, T., Irvin-Williams, E.J., Osborne, F. and Doswell, F. (2011) 'Mobile interfaces for better living: supporting awareness in a smart home environment'. *Proceedings of the 6th international conference on Universal access in human-computer interaction: context diversity – Volume Part III*. Orlando, FL: Springer-Verlag, pp. 163–172.

Graham, S.D.N. (2005) 'Software-sorted geographies'. *Progress in Human Geography*, 29 (5), pp. 562–580.

Grünkranz, D. (2012) *Towards a Phenomenology of Responsive Architecture: Intelligent Technologies and Their Influence on the Experience of Space*. http://www.orambra.com/survey/~phenomenology/media/grunkranz.pdf (Accessed: 1 October 2014).

Hill, D. (2008) 'The Street as Platform.' *City of Sound*. http://www.cityofsound.com/blog/2008/02/the-street-as-p.html (Accessed: 1 October 2014).

Ishii, H., Lakatos, D., Bonanni, L. and Labrune, J.-B. (2012) 'Radical atoms: beyond tangible bits, toward transformable materials'. *interactions*, 19 (1), pp. 38–51.

Ishii, H. and Ullmer, B. (1997) 'Tangible bits: towards seamless interfaces between people, bits and atoms'. *Proceedings of the ACM SIGCHI Conference on Human factors in computing systems*. Atlanta, GA: ACM, pp. 234–241.

Kinsley, S. (2014) 'Memory programmes: the industrial retention of collective life'. *Cultural Geographies*. http://cgj.sagepub.com/content/22/1/155.abstract (Accessed: 31 October 2014).

Kitchin, R. and Dodge, M. (2011) *Code/Space: Software and Everyday Life*. Cambridge, MA: The MIT Press.

Kopytoff, I. (1986) 'The cultural biography of things: commoditization as process', in Appadurai, A. (ed.), *The Social Life of Things: Commodities in Cultural Perspective*. Cambridge: Cambridge University Press, pp. 64–91.

Latour, B. (1987) *Science in Action: How to Follow Scientists and Engineers through Society*. Milton Keynes: Open University Press.

Law, J. (2002) 'Objects and spaces. (Actor-Network Theory)'. *Theory, Culture & Society*, 19 (5/6), pp. 91–105.

Le Corbusier (1989) *Towards a New Architecture*. ed. Etchells, F. Oxford: Butterworth Architecture.

LG (2015) *Discover LG Smart ThinQ™ Washers and Dryers*. http://www.lg.com/us/discover/smartthinq/thinq (Accessed: 1 January 2014).

Lynch, K. (1967) *The Image of the City*. Cambridge, MA: The MIT Press.

Mackenzie, C. (2011) '1993 February: Le Corbusier In the Sun'. *Architectural Review*.

Mayer-Schonberger, V. and Cukier, K. (2013) *Big Data: A Revolution That Will Transform How We Live, Work and Think*. London: John Murray.

McGrenere, J. and Ho, W. (2000) 'Affordances: Clarifying and evolving a concept', *Graphics Interface 2000*. Montreal, Canada, pp. 179–186.

Metz, R. (2013) 'Home Tweet Home: A House with Its Own Voice on Twitter'. *MIT Technology Review*. http://www.technologyreview.com/news/514941/home-tweet-home-a-house-with-its-own-voice-on-twitter/ (Accessed: 1 October 2014).

Miller, D. (2010) *Stuff*. Cambridge: Polity Press.

Moggridge, B. (2007) *Designing Interactions*. Cambridge, MA: The MIT Press.

Negroponte, N. (1995) *Being Digital*. London: Hodder & Stoughton.

Norman, D.A. (1998) *The Design of Everyday Things*. London and Cambridge, MA: The MIT Press.

Picard, R.W. (1997) *Affective Computing*. Cambridge, MA and London: MIT.

Pink, S. and Mackley, K. (2012) 'Video and a sense of the invisible: Approaching domestic energy consumption through the sensory home'. *Sociological Research Online*, 17 (1), p. 3.

Rapoport, A. (1981) 'Identity and Environment: a Cross-cultural Perspective', in Duncan, J.S. (ed.), *Housing and Identity: Cross-cultural Perspectives*. London: Croom Helm.

Rapoport, A. (1982) *The Meaning of the Built Environment: A Nonverbal Communication Approach*. London: Sage Publications.

Rose, D. (2014) *Enchanted Objects: Design, Human Desire, and the Internet of Things*. New York: Scribner Book Company.

Scoble, R. and Israel, S. (2014) *Age of Context: Mobile, Sensors, Data and the Future of Privacy*. USA: Patrick Brewster Press.

Shepard, M. (2010) 'On Hertzian Space and Urban Architecture'. *Vague Terrain*, 16.

Shepard, M. (ed.) (2011) *Sentient City: Ubiquitous Computing, Architecture, and the Future of Urban Space*. Cambridge, MA: The MIT Press.

Steiner, R. (2013) 'The house of the future: London property is kitted out with iPads in the walls, remote-controlled coffee pots, magnetic wallpaper and a 4 METRE widescreen TV'. *The Daily Mail*. http://www.dailymail.co.uk/sciencetech/article-2360496/London-property-kitted-iPads-walls-remote-controlled-coffee-pots-magnetic-wallpaper-4-METRE-widescreen-TV.html, 11 July 2013.

Suchman, L.A. (1987) *Plans and Situated Actions: The Problem of Human-Machine Communication*. Cambridge: Cambridge University Press.

Sung, J.-Y., Guo, L., Grinter, R.E. and Christensen, H.I. (2007) '"My Roomba is Rambo": intimate home appliances'. *Proceedings of the 9th International Conference on Ubiquitous Computing*. Innsbruck, Austria: Springer-Verlag, pp. 145–162.

Turk, V. (2013) 'How to set up a smart house'. *Wired.co.uk.*

Veolia Environment (2013) *Imagine 2050 – Veolia.* http://www.veolia.co.uk/media/research (Accessed: 1 October 2014).

de Waal, M. (2011) 'The Urban Culture of Sentient Cities: From an internet of things to a public sphere centered around things', in Shepard, M. (ed.), *Ubiquitous Computing, Architecture, and the Future of Urban Space.* Cambridge, MA: The MIT Press.

Weiser, M. (1991) 'The Computer for the 21st Century'. *Scientific American*, 265 (3), pp. 94–104.

Wellesley-Miller, S. (1976) 'Intelligent Environments', in Negorponte, N. (ed.), *Soft Architecture Machines.* Cambridge, MA: The MIT Press.

7

Future Challenges

7.1 NETSPACES

In Chapter 6, I described the features and spaces of my analogue house; a far cry from the vision of a connected and sensory 'smart' home. Undoubtedly in the next decade my home environment will transform as radically as the urban spaces documented in the first five chapters of the book. Netspaces will become the norm in many everyday lives and in the urban built environment. The underlying argument in this book has been that space is a framework for how we act in the world, and that this is changed by the emergence of networked technologies and infrastructures. This is based on the premise that we come to know about the world around us and the things in it, through conceptions of space; formed by looking, hearing, touching, imagining, and from description. In doing so we establish relationships between and with things, spaces and people. These relationships are increasingly either with, or mediated by, networked connections and technologies. These changing relationships are being seen at the large scale of infrastructure, in how we communicate and even impact on how time frames our experience of space.

Why not how

In the preceding text I have tried to present a balanced discussion around the changes and developments that are emerging as a result of the intersection and merging of physical and digital space. I have sought to do this by looking at both the positive and negative effects of this merging, and presenting them within a context that draws from both historical events and contemporary real-world case studies. I chose to focus on understanding the nature of these changes, because the fact is that they are happening regardless of whether we explicitly want them or not. This means that the biggest challenge is the one that we face by not engaging with them. The question is therefore, what do we want from networked spaces? This means asking questions. According to Thackara 'we need to ask what purpose will be served by the broadband communications, smart materials, wearable computing, and connected appliances that we're unleashing upon the world.

We need to ask what impact all this stuff will have on our daily lives' (Thackara, 2006, p. 4). Thackara argues that we need to consider design as one of the ways we can address these challenges, and that means making design processes and systems around us intelligible and knowable. In this summary I will reflect on how some of the changing conditions identified in the preceding six chapters address societal challenges raised by the relationship between network technologies and the built environment, and what opportunities remain to be developed. This takes the approach that considers that it is less useful to consider 'how' technology can be used and more important to focus on 'why' and the usefulness of it to people's everyday lives. This approach sees technology as being socially constructed, as opposed to technically determined.

7.2   DESIGN CHALLENGES

Agency

We tend to neglect and even ignore the infrastructure that powers and connects us; we choose not to pay attention to it. From the black box of a Wi-Fi router to the anonymous, industrial-scale shed plastered with air con units or the ubiquitous mobile phone masts; these are generally not considered as part of the city. In Chapter 1, I highlight the 'black boxing' of the, not insubstantial, infrastructures of networks. Whilst disguising a mobile phone mast as a fake tree (as described in Chapter 3), may seem like an idiosyncrasy of the PR sensitive mobile phone industry, in fact it represents a much broader and more pervasive approach to technology in the built space. The 'black boxing' of almost all the technological devices and infrastructures of our networked world occurs at almost all levels; from mobile phone masts to Wi-Fi routers, from building management systems up to the huge scale of data centers. The control and codification of these systems has made them at the very least invisible in our everyday lives and at worst purposely denies us access. These devices and systems emerge only when there is some form of technical failure; the Wi-Fi phone connection goes down or there is a power cut. If we look back in history, many infrastructures are hidden; electricity, water and gas are delivered into our houses and offices and we don't question where they come from; they also require engineers to fix them when they break down. Networked infrastructures and devices include a layer of coding or software to manage the hardware that adds a further dimension to the condition. This brings in to question issues of human agency, or a capacity to act. If we do think about infrastructures more in terms of the 'meshworks' or assemblages that are constituted as a socio-technical formation, then it is important to consider the point at which humans as actors in these networks have the capacity to act. As I highlighted in Chapter 3, in a discussion around digital divides, this is crucial if those who are currently being excluded from network infrastructures can begin to engage and benefit from access to resources and information.

Sassen calls for an 'open source urbanism' (2011b) and for 'talking back to our intelligent city' (2011a) as ways in which to counter the discourse coming from private information technology providers on the responsive ubiquitous networks embedded in the landscape. Sassen argues that this approach could lead 'to a new type of open-source network, where instead of simply having IT workers detect and fix software and code problems as they see them, there would be a collective upgrading and problem-solving dimension involving citizens, a sort of open-source urbanism' (Sassen, 2011a). Easterling, drawing on Latour, introduces the concept of 'active form' to describe a situation where space becomes instrumentalised to do something (2012, p. 188). Latour defines 'doing something' as being about 'making some difference to a state of affairs, transforming some As into Bs through trials with Cs' (2005, pp. 52–53). If we can start to work with infrastructure as composed of different elements, rather than a seamless, invisible, coded system, then we have the potential to work at the boundary between the digital and the material. One approach is through 'hackable' cities (de Waal 2015), where exposing and intervening in the infrastructure is achieved through a bottom-up, citizen led approach.

What we need to remember is that, whilst infrastructure may appear black-boxed and opaque much of what actually lies behind the scenes is in fact messy and incomplete. It suits certain commercial information technology interests to maintain a myth of in-accessibility in order to maintain control over complex systems. For instance in August 2014 it was reported widely that the internet 'had run out of space' (McMillan, 2014). According to a BBC report, a senior analyst at the company Arbor Networks explained that 'this may come as a surprise to non-specialists who view the internet as a high-tech affair comparable to the bridge of the USS Enterprise of Star Trek fame. In actuality, the internet is more akin to an 18th century Royal Navy frigate, with a lot of running about, climbing, shouting, and tugging on ropes required to maintain the desired course and speed' (Ward, 2014). If, instead of assuming that we have no agency in network infrastructures we start to become more viable 'actors', either through technical ability or through how we design environments that reveal and make infrastructures accessible, then we can start to reclaim the space at the boundary between our social and technical worlds.

Community and publics

At the other end of the scale from network infrastructures are social networks; the 'glue' that ties people together. Technology is often blamed for the social isolation particularly the loss of face-to-face contact as people increasingly spend time communicating online or through mobile media (Turkle, 2011). Whilst the literature from as far back as the turn of the twentieth century has claimed that cities can be isolating environments (Simmel, 1950), it is in the last decade that it is asserted that the rise of social media and the increase in time spent online has contributed to a decline in 'social capital'. A similar argument is often made for the apparent privatisation or parochialisation of public space, with mobile phones often seen as being the main culprits. But Putnam, the author of *Bowling Alone* (2000), a key work on the concept of social capital is cautious about the role of online networks in contributing to the decline of community, and instead argues that where online social networks have a physical reality they can support networks of connection and common interest. In Chapter 4, I discussed how the creation of what Varnelis and Ito have termed 'networked publics' (Ito, 2012; Varnelis, 2012) can be seen as a positive counter to the alienation of urban life. Yet, it is only when online connections are linked into to presence in squares, parks and streets that they can create the conditions for 'shared encounters' (Willis et al., 2008) or a performative awareness of others in public space. It is this basic awareness of the presence of others in public space that is often seen as one of the core conditions of a public realm (Arendt, 1999). Work by Wellman, Hampton and colleagues has also shown that online connections can spill offline, and where online and offline connections are not mutually exclusive, they tend to reinforce one another. According to Wellman, the impact of online networks 'on society will be important but evolutionary, like the telephone has been (Fischer, 1992), continuing and intensifying the interpersonal transformation from door-to-door to individualised place-to-place and person-to-person networks. Although face-to-face and telephone contact continue, they are complemented by the Internet's ease in connecting geographically dispersed people and organisations bonded by shared interests' (Wellman et al., 2001, p. 440). As discussed in Chapter 4, movements such as Occupy and the Arab Spring have shown the potential of social activism when online social networks are mobilised in public spaces such as squares and parks. Later in the chapter the potential for urban screens to create 'encounter stages' for the construction of publics was highlighted. In Chapter 3, I focused on the rise of the sharing economy, and how this similarly has the potential to connect people locally through online social networks. Currently the concern is whether these platforms will be appropriated by commercial concerns, who seek to monetise peer-to-peer sharing networks. This reflects a broader challenge of how the characteristics of network structures; where the network tie strength tends to be defined by shared interests and rather than physical proximity, can contribute to social capital at the scale of the neighbourhood or local community.

Resilience

Interestingly one of the most material ways in which networks become materialised is not through human interaction, but through their need for power and energy. Contrary to the myth of lightness and ephemerality, networks actually require significant supplies of energy, and are constructed from material resources that have supply chains. In Chapter 1 I showed how the lightweight metaphor of the cloud is actually a term used to refer to the housing of a large number of globally distributed 'cloud factories' or data centers. The siting of these data centers is increasingly determined by one key concern; the availability of cheap, renewable energy. At the other end of the scale the case study in Chapter 5 of real time cities showed how the provision of 'charging stations' in the transit spaces of our networked world is now a prime concern for those on the move. This represents the emerging of communication and energy networks. If the power supply to your home went down today, so would your internet connection, you would be unable to charge your mobile phone or laptop. For example, if a power cut hit the 'smart' 'APP-artment' featured in Chapter 6, it would rely on switching it's power supply to a local petrol generator, or the smart home home would literally switch off. Where once communication and energy networks were separate, such as in the days of the landline telephone and the coal power station, today they are most intricately interlinked. If we consider the energy challenges of a post-carbon society, then it is not only the challenge of a switch from oil and gas that power our cars and homes that we face, but the fact that the source and size of the energy consumption of our communication networks will also need to change.

Places

In Chapter 2 I discussed the changing nature of space and place, and argued that network infrastructures and connections are not resulting in non-place or placelesness. But they are changing our relationships, experiences and practices that are associated with places. One of the key changes is that it shifts the experience of place from the static, contained and material to one that is inbetween and constantly under construction.

Digital places challenge the dualism of form and function; a space inbetween cannot have a single fixed function attached to it, nor can its boundaries be clearly defined in time or space. In many ways the merging of technological and spatial settings has meant that some of the rules and norms have been detached from the spatial setting and instead are experienced through the technological framework. The software sorting highlighted by Graham (2005) is as much a transfer of existing codes of surveillance onto digital systems, as an introduction of new modes of control. Boundaries in the spatial world are now less defined by territory or ownership but by the ranges of particular technologies or the connective structures of networks. For instance the case study of Wi-Fi in Chapter 3 found that public Wi-Fi creates different spaces of access than the existing public space. When boundaries become more ambiguous then new potentials can emerge. According to Mitchell, when 'the constituent elements of hitherto tightly packaged architectural and urban compositions can begin to float free from one another, and they can potentially relocate and recombine according to new logics' (1995, p. 104). The consequences of this are that are that 'many of our everyday tasks and pastimes will cease to attach themselves to particular spots and slots set aside for their performance-workplaces and working hours, theatres and performance times, home and your own time, and will henceforth be multiplexed and overlaid; we will find ourselves able to switch rapidly from one activity to the other while remaining in the same place, so we will end up using that same place in many different ways' (Mitchell, 1995, p. 100). In this condition hybrid spatial typologies emerge, such as the merging of the spaces of home and work as discussed in Chapter 2. The challenge for society is particularly important to rethink the role and design of buildings that help support community meaning. Libraries, schools, museums and community institutions need to be considered in innovative re-combinations of use that add a new dimension to public space, one that interacts with and supports physical space. A station, such as Paddington, that I described at the start of Chapter 2, supports multiple functions; a place to meet a friend, a temporary office for a skype conference or a backdrop for a mobile film to be posted online. It is revealed to be a platform for a whole range of functions and activities, none of which is made invalid by the original designated function of the railway station. Rather than a space or building containing and defining how it is used, instead the built world becomes transformed into a more flexible backdrop for the performance of everyday life. According to McCullough (2004) one of the consequences of this is that architecture shifts into the background, as an enabler of events, more similar to a stage for the theatre of everyday life than a static and fixed container of functionality.

This raises significant challenges for architects, urban designers and planners, because it is vital to the way we design our urban space. The basis of how we have designed our cities has tended to be as a visual organisation of spatial places connected by infrastructure that connects one place to another based on how close or far away they are. This is now redundant. It is not helpful to maintain the idea that a city is a coherent container with a visible (transparent) structure. It is much more close in organisation to a meshwork or assemblage of relations as described in Chapter 1. As Sassen (2012) points out 'what stands out is that these technologies have not been sufficiently 'urbanised'. On the one hand, cities tend to urbanise technologies – it is not quite feasible to simply plop down a new technology in urban space. This becomes clear, for example, in the fact that the spatial formats through which density is constituted vary sharply across cities; it means that each city partly reshapes even standard technologies' (Sassen, 2012). We need to reinstate boundary objects that interface between spatial and social practices of local and global urban life, and the flows, channels and masts of network infrastructure.

REFERENCES

Arendt, H. (1999) *The Human Condition*. Chicago: University of Chicago Press.

Easterling, K. (2012) 'An Internet of Things'. *E-Flux*.

Graham, S.D.N. (2005) 'Software-sorted geographies'. *Progress in Human Geography*, 29 (5), pp. 562–580.

Ito, M. (2012) 'Introduction', in Varnelis, K. (ed.), *Networked Publics*. Cambridge, MA: The MIT Press.

Latour, B. (2005) *Reassembling the Social: An Introduction to Actor-Network-Theory*. Oxford: Oxford University Press.

McCullough, M. (2004) *Digital Ground: Architecture, Pervasive Computing and Environmental Knowing*. Cambridge, MA: The MIT Press.

McMillan, R. (2014) 'The Internet Has Grown Too Big for Its Aging Infrastructure'. *Wired Magazine.*

Mitchell, W.J. (1995) *City of Bits.* Cambridge, MA: The MIT Press.

Putnam, R.D. (2000) *Bowling Alone: The Collapse and Revival of American Community.* New York and London: Simon & Schuster.

Sassen, S. (2011a) 'Talking back to your intelligent city'. *Voices on Society; McKinsey and Company.* http://voices.mckinseyonsociety.com/talking-back-to-your-intelligent-city/ (Accessed: 1 October 2014).

Sassen, S. (2012) 'Urbanising technology', in Burdett, R. and Rode, P. *The Electric City Newspaper.* http://ec2012.lsecities.net/newspaper/: LSE Cities. 12–14.

Sassen, S. (2011b) 'An interview with Saskia Sassen about "Smart cities"'. http://www.nicolasnova.net/pasta-and-vinegar/2011/07/06/an-interview-with-saskia-sassen-about-smart-cities (Accessed: 13 August 2014).

Simmel, G. (1950) *The Sociology of Georg Simmel,* trans., ed. and introduction by Wolff, K.H., Glencoe, IL Free Press.

Thackara, J. (2006) *In the Bubble: Designing in a Complex World.* Cambridge, MA and London: The MIT Press.

Turkle, S. (2011) *Alone Together: Why We Expect More from Technology and Less from Each Other.* New York: Basic Books.

Varnelis, K. (ed.) (2012) *Networked Publics.* Cambridge, MA: The MIT Press.

Ward, M. (2014) 'Browsing speeds may slow as net hardware bug bites'. *BBC News.* 14 August 2014. http://www.bbc.co.uk/news/technology-28786954 (Accessed: 1 October 2014).

Wellman, B., Haase, A.Q., Witte, J. and Hampton, K. (2001) 'Does the Internet Increase, Decrease, or Supplement Social Capital?: Social Networks, Participation, and Community Commitment'. *American Behavioral Scientist,* 45 (3), pp. 436–455.

Willis, K., Roussos, G., Chorianopoulos, K. and Struppek, M. (2008) *Shared Encounters.* London and New York: Springer.

References

Ads of the World (2012) *Kit Kat: Free no WiFi zone*. http://adsoftheworld.com/media/ambient/kit_kat_free_nowifi_zone (Accessed: 21 August 2014).

AirBnB. *How it Works*. https://www.airbnb.co.uk/help/getting-started/how-it-works (Accessed: 1 October 2014).

Ajuntament de Barcelona (n.n.) *22@Barcelona, the innovation district*. http://www.22barcelona.com/documentacio/22bcn_1T2010_eng.pdf: Ajuntament de Barcelona.

Ajuntament de Barcelona *22@ Infrastructure Plan*. http://81.47.175.201/project-protocol/index.php/22-infrastructure-plan (Accessed: 11 August 2014).

Altman, I. (1975) *The Environment and Social Behaviour*. Monterey, CA: Brooks/Cole.

Anastasi, R., Tandavanitj, N., Flintham, M., Crabtree, A., Adams, M., Row-Farr, J., Iddon, J., Benford, S., Hemmings, T. and Izadi, S. (2002). *Can you see me now? A citywide mixed-reality gaming experience*. http://bit.ly/1OKpU5G.

Andrews, J. and Taylor, J. (1982) *Architecture: A Performing Art*. Oxford: Oxford University Press.

Appadurai, A. (ed.) (1986) *The Social Life of Things: Commodities in Cultural Perspective*. Cambridge: Cambridge University Press.

Arendt, H. (1999) *The Human Condition*. Chicago: University of Chicago Press.

Armbrust, M., Fox, A., Griffith, R., Joseph, A., Katz, R., Kon-winski, A., Lee, G., Patterson, D., Rabkin, A., Stoica, I. and Zaharia, M. (2009) *Above the Clouds: A Berkeley View of Cloud Computing*. Berkeley, CA: University of California, Berkeley.

Arminen, I. and Weilenmann, A. (2009) 'Mobile presence and intimacy – Reshaping social actions in mobile contextual configuration'. *Journal of Pragmatics: An Interdisciplinary Journal of Language Studies*, 41 (10), pp. 1905–1923.

Arquilla, J. and Ronfeldt, D. (2000) *Swarming and the future of conflict*. http://www.rand.org/pubs/documented_briefings/2005/RAND_DB311.pdf: RAND Corporation.

Augé, M. (2009) *Non-places: Introduction to an Anthropology of Supermodernity*. London: Verso Books.

Aurigi, A. (2012) 'Reflections towards an agenda for urban-designing the digital city'. *Urban Design International*, 18 (2), pp. 131–144.

Aurigi, A. and Cindio, F.D. (2008) *Augmented Urban Spaces: Articulating the Physical and Electronic City*. Aldershot: Ashgate.

Auslander, P. (1999) *Liveness: Performance in Mediatized Culture.* London: Routledge.

Ballantyne, A. (2002) *Architecture: a very short introduction.* Oxford and New York: Oxford University Press.

Barcelona free Wifi. http://www.bcn.cat/barcelonawifi/en/ (Accessed: 1 October 2014).

Baskias, H. (2014) 'Pop-up shops spice up airport retail options'. *USA Today.* http://www.usatoday.com/story/travel/flights/2014/02/19/airport-pop-up-stores-retail-shopping/5581917/ (Accessed: 19 February 2014).

BBC (2012) *BBC update on: beyond the broadcast. BBC Outreach Newsletter.* http://downloads.bbc.co.uk/outreach/supplements/FocusOn_PublicPurpose2012.pdf (Accessed: 1 October 2014).

Bell, A.G. (1876) 'Lab notebook, March 10, 1876'. [Notebook] http://www.loc.gov/exhibits/treasures/trr002.html: The Alexander Graham Bell Family Papers.

Benford, S., Crabtree, A., Flintham, M., Drozd, A., Anastasi, R., Paxton, M., Tandavanitj, N., Adams, M. and Row-Farr, J. (2006) 'Can You See Me Now?'. *ACM Transactions on Computer-Human Interaction,* 13 (1), pp. 100–133.

Benjamin, W. (ed.) (2002) *The Arcades Project.* New York: Belknap Press.

Berg Inisght (2014) *Smart Homes and Home Automation.* http://www.berginsight.com/News.aspx.

Berger, W. (1999) 'Lost in Space.' *Wired,* February 1999. http://www.wired.com/1999/02/chiat-3/ (Accessed: 23 October 2015).

Bertelsen, O.W., and Bodker, S. (2001) 'Cooperation in massively distributed information spaces', in Prinz, W., Jarke, M., Rogers, Y., Schmidt, K. and Wulf, V. (eds), *Proceedings of the Seventh European Conference on Computer Supported Cooperative Work,* Bonn, Germany. Dordrecht: Kluwer Academic Publishers, pp. 1–17.

Bertot, J., McClure, C. and Jaeger, P. (2008) 'The impacts of free public Internet access on public library patrons and communities', *Library Quarterly,* 78, pp. 285–301.

Blanchette, J.-F. (2011) 'A material history of bits'. *Journal of the American Society for Information Science and Technology,* 62 (6), pp. 1042–1057.

Blast Theory *Our History & Approach.* www.blasttheory.co.uk/our-history-approach/ (Accessed: 11 October 2014).

Blast Theory *Can You See Me Now?: A game of chase played online and on the streets.* http://www.blasttheory.co.uk/projects/can-you-see-me-now/ (Accessed: 1 January 2014).

Blinkenlights. http://blinkenlights.net/blinkenlights (Accessed: 15 August 2014).

Bond, M. (1958) *Paddington.* London: William Collins & Sons.

Bose, J. and Sharp, J. (2005) 'Measurement of Travel Behavior in a Trip-Based Survey Versus a Time Use Survey: A Comparative Analysis of Travel Estimates Using the 2001 National Household Travel Survey and the 2003 American Time Use Survey'. *ATUS Early Results Conference.* Bethesda, Maryland: 8 and 9 December 2005.

Bourdieu, P. (1992) *The Logic of Practice.* Cambridge: Polity Press.

boyd, d. (2010) 'Social Network Sites as Networked Publics: Affordances, Dynamics, and Implications', in Papacharissi, Z. (ed.), *A Networked Self: Identity, Community, and Culture on Social Network Sites.* Oxford and New York: Routledge, pp. 39–58.

boyd, d. (2011) 'Debating Privacy in a Networked World for the WSJ.' *Techplicy.com*. November 29. http://www.techpolicy.com/Blog/November-2011/Debating-Privacy-in-a-Networked-World-for-the-WSJ.aspx (Accessed: 5 August 2014).

boyd, d. (2014) *It's Complicated: The Social Lives of Networked Teens*. London and New Haven, CT: Yale University Press.

Boyer, C. (1999) 'Crossing CyberCities: Urban Regions and the Cyberspace Matrix', in Beauregard, R. and Body-Gendrot, S. (eds), *The Urban Moment: Cosmopolitan Essays on the Late-20th-Century City*. London: Sage, pp. 51–78.

Boyer, C. (2000) *The City of Collective Memory: Its Historical Imagery and Architectural Entertainments*. Cambridge, MA: The MIT Press.

Brand, S. (2001) interview with Paul Baran. *Wired*.

Bratton, B.H. and Jeremijenko, N. (2008) 'Suspicious Images, Latent Interfaces', The Architectural League of New York. http://www.situatedtechnologies.net/?q=node/88.

Brejzek, T. (2010) 'From social network to urban intervention: On the scenographies of flash mobs and urban swarms'. *International Journal of Performance Arts and Digital Media*, 6 (1), pp. 109–122.

Brown, B., Green, N. and Harper, R. (eds) (2002) *Wireless World: Social, Cultural and Interactional Issues in Mobile Communications and Computing*. London: Springer-Verlag.

Bryant Park Corporation (2014) *Bryant Park Wireless Network*. http://www.bryantpark.org/plan-your-visit/wireless.html (Accessed: 1 October 2014).

Burns, J. (2014) 'Heathrow Installs Sensory Park Pod in T2'. *Airport World*.

Buschauer, R. and Willis, K.S. (eds) (2013) *Locative Media: Multidisciplinary Perspectives on Media and Locality*. Bielefeld: Transcript Verlag.

Cairncross, F. (1997) *The Death of Distance: How the Communications Revolution Will Change Our Lives*. Boston, MA: Harvard Business School Press.

Callon, M. (1986) 'Some Elements of a Sociology of Translation: Domestication of the Scallops and the Fishermen of Saint Brieuc Bay', in Law, J. (ed.), *Power, Action and Belief: A New Sociology of Knowledge?* London: Routledge and Kegan Paul, pp. 196–233.

Carr, N. (2009) *The Big Switch: Rewiring the World, from Edison to Google*. New York: W.W. Norton and Company.

Castells, M. (1996) *The Rise of the Network Society: The Information Age: Economy, Society, and Culture*. Vol. 1. Oxford: Blackwell.

Castells, M. (1998) 'The Education of City Planners in the Information Age'. *Berkeley Planning Journal*, 12 (1), pp. 25–31.

Castells, M. (2004) 'Space of Flows, Space of Places: Materials for a Theory of Urbanism in the Information Age', in Graham, S. (ed.), *The Cybercities Reader*. London: Routledge, pp. 82–93.

Castells, M., Fernández-Ardèvol, M., Qiu, J.L. and Sey, A. (2007) *Mobile Communication and Society: A Global Perspective*. Cambridge and London: The MIT Press.

de Certeau, M. (1984) *The Practice of Everyday Life*. Berkeley, CA: University of California Press.

Chellappa, R. (1997) 'Intermediaries in Cloud-Computing: A New Computing Paradigm'. *INFORMS Annual Meeting*. Dallas, TX: 26–27 October, 1997.

Cisco *Toothpaste, Toilet Paper, and Texting– Say Good Morning to Gen Y*. http://newsroom. cisco.com/press-release-content?articleId=1114955 (Accessed: 7 August 2014).

Citywide IT Services: NYCWiN. http://www.nyc.gov/html/doitt/html/citywide/nycwin.shtml (Accessed: 1 October 2014).

Coates, T. @ *House of Coates Twitter account*. https://twitter.com/houseofcoates (Accessed: 1 October 2014).

Collins, N. (2010) 'Sat nav mistakes: when technology fails'. *The Telegraph*. http://www. telegraph.co.uk/motoring/news/7931558/Sat-nav-mistakes-when-technology-fails. html, 07 Aug 2010.

Cook, G. (2014) *Clicking Clean: How Companies are Creating the Green Internet* http://www. greenpeace.org/usa/Global/usa/planet3/PDFs/clickingclean.pdf: Greenpeace USA.

Copenhagen Airports (2014) *Share your food with the neighbour – a social experiment at Copenhagen Airports*. https://www.cph.dk/en/about-cph/press/news/del-din-mad-med-naboen---socialt-eksperiment-i-kobenhavns-lufthavn-/ (Accessed: 3 October 2014).

Cornflake *Smart App-artment*. http://cornflake.co.uk/ (Accessed: 1 January 2015).

Cosgrove, D. (1996) 'Windows on the City'. *Urban Studies*, 33 (8), pp. 1495–1498.

Coyne, R. (2010) *The Tuning of Place: Sociable Spaces and Pervasive Digital Media*. Cambridge, MA: The MIT Press.

Crabtree, A., Benford, S., Rodden, T., Greenhalgh, C., Flintham, M., Anastasi, R., Drozd, A., Adams, M., Row-Farr, J., Tandavanitj, N. and Steed, A. (2004) 'Orchestrating a mixed reality game "on the ground"'. *SIGCHI Conference on Human Factors in Computing Systems (CHI '04)*. Vienna, Austria: 24–29 April 2004 ACM, pp. 391–398.

Cramer, H., Rost, M. and Holmquist, L.E. (2011) 'Performing a check-in: emerging practices, norms and "conflicts" in location-sharing using foursquare'. *13th International Conference on Human Computer Interaction with Mobile Devices and Services (MobileHCI '11)*. New York: ACM, pp. 57–66.

Crang, M., Crosbie, T. and Graham, S. (2007) 'Technology, timespace and the remediation of neighbourhood life'. *Environment and Planning A*, 39 (10), pp. 2405–2422.

Crang, M. and Graham, S. (2007) 'Sentient cities: ambient intelligence and the politics of urban space'. *Information, communication society*, 10 (6), pp. 789–781.

Croft, J. (n.d.) *An FAQ on the LETS system*. http://www.gdrc.org/icm/lets-faq.html (Accessed: 1 October 2014).

Crow, B. and Miller, T. (2006) *Community Wireless Infrastructure Research Project1: Île Sans Fil Case Study Map*. http://www.cwirp.ca/files/CWIRP_ISF_case.pdf: Ryerson University.

Daniels, P.W. (1975) *Office Location*. London: Bell and Sons.

Davenport, T.H. and Beck, J.C. (2001) *The Attention Economy: Understanding the New Currency of Business*. Boston, MA: Harvard Business School Press.

Dayan, D. and Katz, E. (1992) *Media Events: The Live Broadcasting of History*. Cambridge, MA: Harvard University Press.

Dodge, M. and Kitchen, R. (2004) 'Flying through code/space: the real virtuality of air travel'. *Environment and Planning A*, 36 (2), pp. 195–211.

Dourish, P. (2001) *Where the Action Is: The Foundations of Embodied Interaction*. Cambridge, MA and London: The MIT Press.

Dourish, P. and Mazmanian, M. (2011) 'Media as material: Information representations as material foundations for organizational practice'. *Third International Symposium on Process Organization Studies*. Corfu, Greece: 16–18 June 2011.

Draadloos Groningen. (2015). http://draadloosgroningen.nl/wordpress/ (Accessed: 1 October 2014).

Drucker, J. (2013) 'Google Joins Apple Avoiding Taxes With Stateless Income'. *Bloomberg*. http://www.bloomberg.com/news/2013-05-22/google-joins-apple-avoiding-taxes-with-stateless-income.html, 22 May 2013.

Dunne, A. (1999) *Hertzian Tales: Electronic Products, Aesthetic Experience and Critical Design*. London: RCA CRD Research Publications.

Dwyer, C. and Jackson, P. (2003) 'Commodifying difference: selling EASTern fashion'. *Environment and Planning D: Society and Space*, 21, pp. 269–291.

Eagle, N. and Pentland, A. (2006) 'Reality mining: sensing complex social systems'. *Personal Ubiquitous Computing*, 10 (4), pp. 255–268.

Easterling, K. (2012) 'An Internet of Things'. *E-Flux*.

Echikson, W. (2014) 'Expanding our data centres in Europe'.

Edensor, T. (ed.) (2010) *Geographies of Rhythm*. Aldershot: Ashgate.

Ewalt, D.M. (2013) 'America's Most Crowded Airports'. *Forbes*.

Fagen, M.D. (1975) *A History of Engineering & Science in the Bell System, The Early Years (1875–1925)*. Bell Telephone Laboratories, Inc.

Farias, I. (2011) 'An interview with Stephen Graham', in Farias, I. and Bender, T. (eds), *Urban Assemblages: How Actor-Network Theory Changes Urban Studies.* Abingdon: Routledge, pp. 197–206.

Feenberg, A. (1999) *Questioning technology.* London: Routledge.

Ferdinando, L. (2012) *FAA Expects Travel on US Airlines to Nearly Double in 20 Years* Voice of America. 7 March 2012. Available at: http://www.voanews.com/content/faa-expects-travel-on-us-airlines-to-nearly-double-in-20-years-141988063/179138.html.

Feuer, A. (2012) 'Occupy Sandy: A Movement Moves to Relief'. *The New York Times*. http://www.nytimes.com/2012/11/11/nyregion/where-fema-fell-short-occupy-sandy-was-there.html?pagewanted=all&_r=0, 9 November 2012.

Folger, T. (2014). 'Revealed World'. *National Geographic*. http://ngm.nationalgeographic.com/big-idea/14/augmented-reality-pg2.

Forlano, L. (2009) 'WiFi Geographies: When Code Meets Place'. *Inf. Soc.*, 25 (5), pp. 344–352.

Forlano, L. (2013) 'Making waves; Urban Technology and the co-production of place'. *First Monday*, 18 (11).

Foursquare (2011) *10 Million*. https://foursquare.com/10million (Accessed: 1 October 2012).

Foursquare (2014a). www.foursquare.com (Accessed: 14 August 2014).

Foursquare (2014b) 'Our crowd-sourced places database has over 60,000,000 entries and 5,000,000,000 check-ins, and one major new partner – Microsoft', in *The Foursquare Blog*. http://blog.foursquare.com/post/75603461066/our-crowd-sourced-places-database-has-over: 2014 (Accessed: 14 August 2014).

Frith, J. (2012) *Constructing Location, One Check-in at a Time: Examining the Practices of Foursquare Users*. NC State University.

Fujimoto, K. (2005) 'The Third-Stage Paradigm: Territory Machines from the Girls' Pager Revolution to Mobile Aesthetics', in Ito, M., Okabe D. and Matsuda, M. (eds), *Personal, Portable, Pedestrian: Mobile Phones in Japanese Life*. Cambridge, MA: The MIT Press.

Furuto, A. (2011) '"Urban Future" at Design Miami 2011 / BIG + Kollision + Schmidhuber & Partner'. *Arch Daily*, 2014 (21 August).

Galloway, K. and S. Rabinowitz (1980). *Hole in Space*. http://www.ecafe.com/getty/HIS/ (Accessed: 1 October 2014).

Garfinkel, S. (2011) 'The Cloud Imperative. Business Report.' *MIT Technology Review*. http://www.technologyreview.com/news/425623/the-cloud-imperative/ (Accessed: 1 October 2014).

Gaver, W.W. (1991) 'Technology affordances'. *Proceedings of the SIGCHI Conference on Human Factors in Computing Systems*. New Orleans, Louisiana, USA: ACM, pp. 79–84.

Gehl, J. (1987) *Life Between Buildings: Using Public Space*. New York: Van Nostrand Reinhold.

Gehl, J. (2011) *Life between buildings* ? Washington, DC: Island Press.

Geng, H. (2015) *Data Center Handbook*. Hoboken, NJ: John Wiley and Son.

Gergen, K.J. (2002) 'The challenge of absent presence', in Katz, J.E. and Aakhus, M. (eds), *Perpetual Contact: Mobile Communication, Private Talk, Public Performance*. Cambridge: Cambridge University Press, pp. 227–241.

Getaround *How it Works*. https://www.getaround.com/tour (Accessed: 1 January 2014).

Gibson, J. (1979) *The Ecological Approach to Visual Perception*. Boston, MA: Houghton Mifflin.

Gibson, W. (1995) *Neuromamcer*. London: Harper Voyager.

Glanz, J. (2012) 'The Cloud Factories: Data Barns in a Farm Town, Gobbling Power and Flexing Muscle'. *The New York Times*. http://www.nytimes.com/2012/09/24/technology/data-centers-in-rural-washington-state-gobble-power.html?pagewanted=all, 23 September 2012.

Goffman, E. (1963) *Behaviour in Public Places: Notes on the Social Organization of Gatherings*. New York: Free Press.

Goffman, E. (1983) 'The Interaction Order'. *American Sociological Review*, 48 (1), pp. 1–17.

Goffman, E. (1990) *The Presentation of Self in Everyday Life*. London: Penguin.

Goldwyn, E. (2014) 'Will Uber Destroy the Driving Profession?' *The New Yorker*. http://www.newyorker.com/tech/elements/will-uber-destroy-the-driving-profession (Accessed: 8 August 2014).

Gomez, R. and Camacho, K. (2011) 'Who Uses Public Access Venues?', in Gomez, R. and Camacho, K. (eds), *Libraries, Telecentres, Cybercafes and Public Access to ICT: International Comparisons*. Hershey, PA: IGI Global, pp. 11–22.

Google (2012) *Our Mobile Planet: Global Smartphone Users*. http://goo.gl/WAFg7: Google.

Google (2014) *Loon for all: Balloon-powered internet for everyone*. http://www.google.com/loon/ (Accessed: 1 October 2014).

Google Data Centers *Hamina, Finland*. http://www.google.co.uk/about/datacenters/inside/locations/hamina/ (Accessed: 1 October 2014).

Google Data Centers *Locations*. http://www.google.co.uk/about/datacenters/inside/locations/ (Accessed: 1 January 2014).

Gordon, E. and de Souza e Silva, A. (2011) *Net Locality: Why Location Matters in a Networked World*. Chichester, UK: Wiley-Blackwell.

Gottdiener, M. (2001) *Life in the Air: Surviving the New Culture of Air Travel*. Lanham, MA: Rowman and Littlefield.

Gottmann, J. (1977) 'Megalopolis and Antipolis: The Telephone and the Structure of the City', in de Sola Pool, I. (ed.), *The Social Impact of the Telephone*. Cambridge, MA: The MIT Press, pp. 303–317.

GPS Bites (2012) 'The Top 10 List of Worst GPS Disasters and Sat Nav Mistakes'. *GPS Bites*. http://www.gpsbites.com/top-10-list-of-worst-gps-disasters-and-sat-nav-mistakes (Accessed: 1 October 2014).

Gracanin, D., McCrickard, D.S., Billingsley, A., Cooper, R., Gatling, T., Irvin-Williams, E.J., Osborne, F. and Doswell, F. (2011) 'Mobile interfaces for better living: supporting awareness in a smart home environment'. *Proceedings of the 6th international conference on Universal access in human-computer interaction: context diversity – Volume Part III*. Orlando, FL: Springer-Verlag, pp. 163–172.

Graham, S. (1997) 'Cities in the real-time age: the paradigm challenge of telecommunications to the conception and planning of urban space'. *Environment and Planning A*, 29 (1), pp. 105–127.

Graham, S. and Marvin, S. (1996) *Telecommunications and the City*. London: Routledge.

Graham, S. and Marvin, S. (2001) *Splintering Urbanism: Networked Infrastructures, Technological Mobilities and the Urban Condition*. London: Routledge.

Graham, S.D.N. (2005) 'Software-sorted geographies'. *Progress in Human Geography*, 29 (5), pp. 562–580.

Green, N. (2002) 'On the move: Technology, mobility, and the mediation of social time and space'. *Inf. Soc.*, 18 (4), pp. 281–292.

Greenfield, A. (2006) *Everyware: The Dawning Age of Ubiquitous Computing*. Berkeley, CA: New Riders.

Gregoire, C. (2013) *How Technology Speeds Up Time (And How To Slow It Down Again)*. *The Huffington Post*. 12/06/2013. Available at: http://www.huffingtonpost.com/2013/12/06/technology-time-perception_n_4378010.html.

Grünkranz, D. (2012) *Towards a Phenomenology of Responsive Architecture: Intelligent Technologies and Their Influence on the Experience of Space*. http://www.orambra.com/survey/~phenomenology/media/grunkranz.pdf (Accessed: 1 October 2014).

Habermas, J., Lennox, S. and Lennox, F. (1974) 'The Public Sphere: An Encyclopedia Article (1964)'. *New German Critique* (3), pp. 49–55.

Habermas, J. (1989). 'The Public Sphere: An Encyclopedia Article', in Bronner, S.E. and Kellner, D. (eds), *Critical Theory and Society: A Reader*. New York: Routledge, pp. 136–142.

Habermas, J. (1999) *The Structural Transformation of the Public Sphere: An Enquiry into a Category of Bourgeois Society*. Oxford: Polity.

Habuchi, I. (2005) 'Accelerating Reflexivity', in Ito, M., Okabe, D. and Matsuda, M. (eds), *Personal, Portable, Pedestrian: Mobile Phones in Japanese Life*. Cambridge, MA: The MIT Press.

Haddon, L., de Gournay, C., Lohan, M., Östlund, B., Palombini, I. and Sapio, B. (2003) *From Mobile to Mobility: The Consumption of ICTs and. Mobility in Everyday Life*. http://www.lse.ac.uk/media@lse/whosWho/…/Mobility and ICTs.pdf.

Hall, E. (1966) *The Hidden Dimension*. Garden City, NY: Doubleday Anchor Books.

Hampton, K. (2007) 'Neighborhoods in the Network Society: The e-Neighbors Study'. *Information, Communication & Society*, 10 (5), pp. 714–748.

Hampton, K., Livio, O. and Sessions, L. (2010) 'The Social Life of Wireless Urban Spaces: Internet Use, Social Networks, and the Public Realm'. *Journal of Communication*, 60 (4), pp. 701–722.

Hannam, K., Sheller, M. and Urry, J. (2006) 'Editorial: mobilities, immobilities and moorings'. *Mobilities*, 1 (1), pp. 1–22.

Harrison, S. and Dourish, P. (1996) 'Re-Place-ing Space: The Roles of Space and Place in Collaborative Systems'. *ACM Conference Computer-Supported Cooperative Work CSCW'96* Boston, MA: ACM, pp. 67–76.

Harrison, T. and Stephen, T. (1999) 'Researching and Creating Community Networks', in Jones, S. (ed.), *Doing Internet Research: Critical Issues and Methods for Examining the Net* London: Sage Publications.

Harvey, D. (1989) *The Urban Experience*. Oxford: Blackwell.

Harvey, D. (2006) 'The Political Economy of Public Space', in Low, S. and Smith, N. (eds), *The Politics of Public Space*. New York: Routledge.

Harvey, D. (2012) *Rebel Cities: From the Right to the City to the Urban Revolution*. London: Verso.

Heathrow Airport *Phones – public, rental and charging*. http://www.heathrowairport.com/heathrow-airport-guide/services-and-facilities/phones-and-charging (Accessed: 1 October 2014).

Herman, R. and Ausubel, J. (1988) *Cities and Their Vital Systems: Infrastructure Past Present and Future*. Washington, DC: National Academy Press.

Hight, J. (2013) 'Narrative archealogy', in Buschauer, R. and Willis, K.S. (eds), *Locative Media: Multidisciplinary Perspectives on Media and Locality*. Bielefeld: Transcript Verlag.

Hill, D. (2008) 'The Street as Platform.' *City of Sound*. http://www.cityofsound.com/blog/2008/02/the-street-as-p.html (Accessed: 1 October 2014).

Hill, D. (2012). 'In praise of lost time.' *Domus*. http://www.domusweb.it/en/design/2012/03/05/in-praise-of-lost-time.html (Retrieved: 1 August 2014).

Hillier, B. and Hanson, J. (1984) *The Social Logic of Space*. Cambridge: Cambridge University Press.

Höflich, J. (2005) 'The Mobile Phone and the Dynamic between Private and Public Communication: Results of an International Exploratory Study', in Glotz, P., Bertschi, S. and Locke, C. (eds), *Thumb Culture: The Meaning of Mobile Phones in Society*. Bielefeld: Transcript, pp. 123–136.

Höflich, J.R. (2006) 'The mobile phone and the dynamic between private and public communication: Results of an international exploratory study'. *Knowledge, Technology & Policy*, 19 (2), pp. 58–68.

Horan, T. (2000) *Digital Places: Building Our City of Bits*. Washington, DC: Urban Land Institute.

Hornecker, E., Marshall, P. and Rogers, Y. (2007) 'From entry to access: how shareability comes about'. *Proceedings of the 2007 conference on Designing pleasurable products and interfaces*. Helsinki, Finland: ACM, pp. 328–342.

Humphreys, L. and Liao, T. (2013) 'Foursquare and the Parochialization of Public Space'. *First Monday*, 18 (11).

Hutchins, E. (1996) *Cognition in the Wild*. Cambridge, MA: The MIT Press.

Hyman, I., Boss, M., Wise, B., McKenzie, K. and Caggiano, J. (2010) 'Did you see the unicycling clown? Inattentional blindness while walking and talking on a cell phone'. *Applied Cognitive Psychology*, 24 pp. 597–607.

Impact Hubs (2014) *Impact Hubs*. http://www.impacthub.net/what-is-impact-hub (Accessed: 21 August 2014).

International Urban Screens Association (n.n.) *about urban screens: about*. International Urban Screens Association. http://www.urbanscreensassoc.org/about/ (Accessed: 1 October 2014).

Ishii, H., Lakatos, D., Bonanni, L. and Labrune, J.-B. (2012) 'Radical atoms: beyond tangible bits, toward transformable materials'. *interactions*, 19 (1), pp. 38–51.

Ishii, H. and Ullmer, B. (1997) 'Tangible bits: towards seamless interfaces between people, bits and atoms'. *Proceedings of the ACM SIGCHI Conference on Human factors in computing systems*. Atlanta, GA: ACM, pp. 234–241.

Ito, M. (2012) 'Introduction', in Varnelis, K. (ed.), *Networked Publics*. Cambridge, MA: The MIT Press.

Ito, M. and Okabe, D. (2005) 'Technosocial situations: emergent structuring of mobile e-mail use', in Ito, M., Okabe, D. and Matsuda, M. (eds), *Personal, Portable, Pedestrian: Mobile Phones in Japanese Life*. Cambridge, MA: The MIT Press.

Ito, M., Okabe, D. and Andersen, K. (2010) 'Portable Objects in Three Global Cities: The Personalization of Urban Places', in Ling, R. and Campbell, S.W. (eds), *The Reconstruction of Space and Time: Mobile Communication Practices*. New Brunswick, NJ: Transaction Publishers, pp. 67–87.

Iveson, K. (2009) 'Too Public or Too Private? The Politics of Privacy in the Real-time City'. *Engaging Data Forum*. Cambridge, MA, 13 October 2009.

Jacobs, J. (2002) *The Death and Life of Great American Cities*. New York: Random House.

Jeffries, A. (2013) 'Taxi race: can Uber and Hailo deliver a real-time revolution?' *The Verge* http://www.theverge.com/2013/2/7/3964394/taxi-race-can-uber-and-hailo-deliver-a-real-time-revolution (Accessed: 6 August 2014).

Katz, J.E. and Aakhus, M. (eds) (2002) *Perpetual Contact: Mobile Communication, Private Talk, Public Performance*. Cambridge: Cambridge University Press.

Kelleher, K. (2013) 'Mobile growth is about to be staggering'. *Fortune*. http://fortune.com/2013/02/20/mobile-growth-is-about-to-be-staggering/ (Accessed: 23 October 2015).

Kern, S. (1983) *The Culture of Time and Space, 1880–1918*. Cambridge MA: Harvard University Press.

Kinsley, S. (2014) 'Memory programmes: the industrial retention of collective life'. *Cultural Geographies*. http://cgj.sagepub.com/content/22/1/155.abstract (Accessed: 31 October 2014).

Kitchin, R. and Dodge, M. (2011) *Code/Space: Software and Everyday Life*. Cambridge, MA: The MIT Press.

Knorr-Cetina, K. (2003) 'From Pipes to Scopes: The Flow Architecture of Financial Markets'. *Distinktion*, 7 pp. 7–23.

Kolarevic, B. and Malkawi, A.M. (2005) *Performative Architecture: Beyond Instrumentality*. New York: Spon Press.

Koomey, J. (2011) *Growth in data center electricity use 2005 to 2010*. Oakland, CA. Available at: http://www.analyticspress.com/datacenters.html.

Kopytoff, I. (1986) 'The cultural biography of things: commoditization as process', in Appadurai, A. (ed.), *The Social Life of Things: Commodities in Cultural Perspective*. Cambridge: Cambridge University Press, pp. 64–91.

Krempl, S. (2002). *Public Spaces Invaders*. http://www.heise.de/tp/artikel/11/11798/1.html (Accessed: 1 October 2014).

Lasen, A. (2003) 'A comparative study of mobile phone use in public places in London, Madrid and Paris'.

Lash, S. and Urry, J. (1994) *Economies of Signs and Space*. London: Sage.

Latchem, C. and Walker, D. (2001) *Telecentres: Case Studies and Key Issues*. Vancouver: The Commonwealth of Learning.

Latour, B. (1987) *Science in Action: How to Follow Scientists and Engineers through Society*. Milton Keynes: Open University Press.

Latour, B. (1993) *We Have Never Been Modern*. London: Harvester Wheatsheaf.

Latour, B. (2005) *Reassembling the Social: An Introduction to Actor-Network-Theory*. Oxford: Oxford University Press.

Latour, B. (2011) 'Networks, Societies, Spheres: Reflections of an Actor-Network Theories'. *International Journal of Communication*, 5 pp. 796–810.

Latour, B. and Hermant, E. (2004) *Paris: invisible city*. http://www.bruno-latour.fr/virtual/EN/index.html (Accessed: 1 October 2014).

Latour, B. and Yaneva, A. (2008) 'Give Me a Gun and I will Make All Buildings Move: An ANT's View of Architecture', in Geiser, R. (ed.), *Explorations in Architecture: Teaching, Design, Research*. Basel: Birkhäuser, pp. 80–89.

Law, J. (1994) *Organizing Modernity: Social Order and Social Theory*. Oxford: Blackwell.

Law, J. (2002) 'Objects and spaces. (Actor-Network Theory)'. *Theory, Culture & Society*, 19 (5/6), pp. 91–105.

Lawson, B. (2005) *The Language of Space*. Oxford: Architectural Press.

Le Corbusier (1989) *Towards a New Architecture*. ed. Etchells, F., Oxford: Butterworth Architecture.

Lefebvre, H. (1991) *The Production of Space*. Oxford and Cambridge, MA: Blackwell.

Lefebvre, H. (2004) *Rhythmanalysis: Space, Time and Everyday Life*. Translated by Stuart Elden and Gerald Moore. London, New York: Continuum.

Leon, N. (2008) 'Attract and connect: The Barcelona innovation district and the internationalization of Barcelona business'. *Innovation: management, policy, and practice*, 10, pp. 235–246.

Lewis, S., Pea, R. and Rosen, J. (2010) 'Collaboration with Mobile Media: Shifting from "Participation" to "Co-creation"', *6th IEEE International Conference on Wireless, Mobile and Ubiquitous Technologies in Education*. Kaohsiung, Taiwan IEEE Computer Society, pp. 112–116.

LG (2015) *Discover LG Smart ThinQ™ Washers and Dryers*. http://www.lg.com/us/discover/smartthinq/thinq (Accessed: 1 January 2014).

Licoppe, C. (2004) "'Connected' presence: the emergence of a new repertoire for managing social relationships in a changing communication technoscape'. *Environment and Planning D: Society and Space*, 22 (1), pp. 135–156.

Licoppe, C. (2010) 'The "Crisis of the Summons": A Transformation in the Pragmatics of "Notifications," from Phone Rings to Instant Messaging'. *The Information Society*, 26 (4), pp. 288–302.

Ling, R. (1997) *'One Can Talk about Common Manners!': The Use of Mobile Telephones in Inappropriate Situations.* Stokholm: Telia.

Ling, R. (2005) *The Mobile Connection: The Cell Phone's Impact on Society.* San Francisco, CA and Oxford: Elsevier/Morgan Kaufmann.

Ling, R. and Yttri, B. (2002) 'Hyper-coordination via mobile phones in Norway', in Katz, J. and Aakhus, M. (eds), *Perpetual Contact: Mobile Communication, Private Talk, Public Performance.* Cambridge: Cambridge University Press, pp. 139–169.

Lofland, L.H. (1998) *The Public Realm: Exploring the City's Quintessential Social Territory.* Hawthorne, NY: Aldine de Gruyter.

Low, S. and Smith, N. (eds) (2006) *The Politics of Public Space.* New York: Routledge.

Lynch, K. (1967) *The Image of the City.* Cambridge, MA: The MIT Press.

Lynch, K. (1988) *What Time is the Place?* Cambridge, MA: The MIT Press.

Lyster, C. (2013) 'The Future of Mobility: Greening the Airport'. *Places Journal*, July 2013.

Mackenzie, A. (2008) 'FCJ-085 Wirelessness as Experience of Transition'. *FibreCulture* (13).

Mackenzie, A. (2010) *Wirelessness.* Cambridge, MA: The MIT Press.

Mackenzie, C. (2011) '1993 February: Le Corbusier In the Sun'. *Architectural Review*.

Macleod, I. (2013) '89% browse or buy using smartphones and tablets during their commute'. *The Drum.* http://www.thedrum.com/news/2013/12/04/89-browse-or-buy-using-smartphones-and-tablets-during-their-commute (Accessed: 25 August 2014).

Manovich, L. (2006) 'The poetics of augmented space'. *Visual Communication*, 5 (2), pp. 219–240.

Manzo, L. and Perkins, D. (2006) 'Finding Common Ground: The Importance of Place Attachment to Community Participation and Planning'. *Journal of Planning Literature*, 20 pp. 335–350.

Marchetti, C. (1994) *Anthropological Invariants in Travel Behavior, Technological Forecasting and Social Change* http://www.cesaremarchetti.org/archive/electronic/basic_instincts.pdf: International Institute for Applied Systems Analysis, Laxenburg, Austria, pp. 75–88.

MarketsandMarkets (2014) *Cloud Computing Market: Global Forecast (2010–2015).* http://www.marketsandmarkets.com/Market-Reports/cloud-computing-234.html: PR Web.

Marvin, S. and Graham, S. (1994) 'Privatization of utilities: the implications for cities in the United Kingdom'. *Journal of Urban Technology*, 2 (1), pp. 47–66.

Massey, D. (2005) *For Space.* London: Sage.

Massey, J. and Snyder, B. (2012) 'Occupying Wall Street: Places and Spaces of Political Action'. *Places Journal*.

Mawoud, J. (2014) 'Dubai, Once a Humble Refueling Stop, Is Crossroad to the Globe'. *The New York Times.* http://www.nytimes.com/2014/06/19/business/international/once-a-humble-refueling-stop-dubai-is-crossroad-to-the-globe.html?_r=2, JUNE 18, 2014.

Mayer-Schonberger, V. (2009) *Delete: The Virtue of Forgetting in the Digital Age*. Princeton, NJ: Princeton University Press.

Mayer-Schonberger, V. and Cukier, K. (2013) *Big Data: A Revolution That Will Transform How We Live, Work and Think*. London: John Murray.

McCullough, M. (2004) *Digital Ground: Architecture, Pervasive Computing and Environmental Knowing*. Cambridge, MA: The MIT Press.

McCullough, M. (2014) *Ambient Commons: Attention in the Age of Embodied Information*. Cambridge, MA: The MIT Press.

McGrenere, J. and Ho, W. (2000) 'Affordances: Clarifying and evolving a concept', *Graphics Interface 2000*. Montreal, Canada, pp. 179–186.

McLaren, C. (2011) 'The Senseable City: An Interview with Carlo Ratti', in *LabLog*. http://blogs.guggenheim.org/lablog/the-senseable-city-an-interview-with-carlo-ratti/: Guggenheim. 2014.

Mcluhan, M. (1964) *Understanding Media: The Extensions of Man*. New York: McGraw-Hill.

McMillan, R. (2014) 'The Internet Has Grown Too Big for Its Aging Infrastructure'. *Wired Magazine*.

McQuire, S. (2006) 'The politics of public space in the media city'. *First Monday*, 4 (Special Issue #4: Urban Screens: Discovering the potential of outdoor screens for urban society).

McQuire, S. (2010) 'Rethinking media events – large screens, public space broadcasting and beyond'. *New Media & Society*, 12 (4), pp. 567–582.

Mels, T. (ed.) (2004) *Reanimating Places: A Geography of Rhythms*. Aldershot: Ashgate.

Metz, R. (2013) 'Home Tweet Home: A House with Its Own Voice on Twitter'. *MIT Technology Review*. http://www.technologyreview.com/news/514941/home-tweet-home-a-house-with-its-own-voice-on-twitter/ (Accessed: 1 October 2014).

Meyrowitz, J. (1985) *No Sense of Place: The Impact of Electronic Media on Social Behavior*. New York: Oxford University Press.

Miller, D. (2010) *Stuff*. Cambridge: Polity Press.

Miller, R. (2014) 'Will the Netflix Model Gain Traction? Why Service Providers Should Take Note'. *Data Center Knowledge*. http://www.datacenterknowledge.com/archives/2014/01/07/vc-firm-hires-netflix-infrastructure-guru-will-invest-cloud-enablers/ (Accessed: 1 October 2014).

Minute Suites. www.minutesuites.com (Accessed: 1 October 2014).

MIT SENSEable City Lab (2006) *Real Time Rome*. http://senseable.mit.edu/realtimerome/ (Accessed: 1 October 2014).

Mitchell, W.J. (1995) *City of Bits*. Cambridge, MA: The MIT Press.

Mitchell, W.J. (2002) 'E-Bodies, E-Building, E-Cities', in Leach, N. (ed.), *Designing for a Digital World*. London: Wiley Academic, pp. 50–56.

Mitchell, W.J. (2004) *Me++: The Cyborg Self and the Networked City*. Cambridge, MA: The MIT Press.

Mobile in the Airtravel Industry Report 2014 (2014) http://www.eyefortravel.com/mobile-and-technology/mobile-airtravel-industry-report-2014: Eye For Travel.

Moggridge, B. (2007) *Designing Interactions*. Cambridge, MA: The MIT Press.

Moores, S. (2000) *Media and Everyday Life in Modern Society*. Edinburgh: Edinburgh University Press.

Morgenland. http://www.morgenland-berlin.de/ (Accessed: 1 October 2014).

Morris, S. and Gibson, O. (2006) 'Blow to BBC image as Liverpool and London pull the plug on big screens'. *The Guardian*. http://www.theguardian.com/media/2006/jun/13/broadcasting.bbc1, Tuesday 13 June 2006.

Moss, M. and Townsend, A. (1999) 'How telecommunications systems are transforming urban spaces', in Wheeler, J.O. and Aoyama, Y. (eds), *Fractured Geographies: Cities in the Telecommunications Age*. New York: Routledge.

Mumford, L. (1934) *Technics and Civilization*. San Diego, CA: Harcourt, Brace and Company.

Mumford, L. (1961) *The City in History*. New York: Harcourt Brace.

Muzychenko, A. and Kats, P. (2015) *Most popular Foursquare venues*. Foursquare Stats. http://www.4sqstat.com/ (Accessed: 1 January 2015).

n.n. (1922) 'Radio; Farm Service Growing'. *The New York Times*. http://query.nytimes.com/gst/abstract.html?res=990CE0D91239EF3ABC4850DFB1668389639EDE. (6), 30 July 1922.

Needleman, R. (2009) 'Starbucks: Stay as long as you want'. *CNet*. http://www.cnet.com/uk/news/starbucks-stay-as-long-as-you-want/ (Accessed: 1 October 2014).

Negroponte, N. (1995) *Being Digital*. London: Hodder & Stoughton.

Norman, D.A. (1998) *The Design of Everyday Things*. London: The MIT Press.

Nye, D.E. (1990) *Electrifying America: Social Meanings of a New Technology*. Cambridge, MA: The MIT Press.

O'Hara, K., Glancy, M. and Robertshaw, S. (2008) 'Understanding collective play in an urban screen game', *2008 ACM conference on Computer supported cooperative work (CSCW '08)*. New York: ACM, pp. 67–76.

Offenhuber, D. and Ratti, C. (2012) 'Reading the City – Reconsidering Kevin Lynch's Notion of Legibility in the Digital Age', in Berzina, Z., Junge, B., Westerveld, W. and Zwick, C. (eds), *The Digital Turn – Design in the Era of Interactive Technologies*. Zürich: Park Books.

Oldenburg, R. (1999) *The Great Good Place: Cafés, Coffee Shops, Bookstores, Bars, Hair Salons, and Other Hangouts at the Heart of a Community*. New York: Marlowe & Company.

Pascoe, C. (2011) 'Time, technology and leaping seconds', http://googleblog.blogspot.de/2011/09/time-technology-and-leaping-seconds.html (Accessed: 2014).

Pedersen, I. (2009) 'Radiating Centers: Augmented Reality and Human-Centric Designs', *IEEE International Symposium on Mixed and Augmented Reality*. Orlando, FL, 19–22 October 2009, pp. 11–16.

Pertierra, R. (2005) 'If You Can't Afford a Room of Your Own, Buy a Mobile Phone'. *Conference on Mobile Communication and Asian Modernities*. Hong Kong, SAR: 8 June 2005.

Pew Internet Project (2014) 'Social Networking Fact Sheet'. [Online]. http://www.pewinternet.org/fact-sheets/social-networking-fact-sheet/.

Picard, R.W. (1997) *Affective computing*. Cambridge, MA and London: The MIT Press.

Pink, S. and Mackley, K. (2012) 'Video and a sense of the invisible: Approaching domestic energy consumption through the sensory home'. *Sociological Research Online*, 17 (1), pp. 3.

de Sola Pool, I. (1977) *The Social Impact of the Telephone*. Cambridge, MA: The MIT Press.

de Sola Pool, I. (1998) *Politics in Wired Nations*. New York: Transaction Publishers.

de Sola Pool, I. Decker, C., Lizard, S., Israel, K., Rubin, P. and Weinstein, B. (1977) 'Foresight and Hindsight: The Case of the Telephone', in de Sola Pool, I. (ed.), *The Social Impact of the Telephone*. Cambridge, MA: The MIT Press, pp. 127–157.

Popper, B. and Hamburger, E. (2014). 'Meet Swarm: Foursquare's ambitious plan to split its app in two: To take on Yelp, Foursquare is moving beyond the check-in.' *The Verge*. http://www.theverge.com/2014/5/1/5666062/foursquare-swarm-new-app (Accessed: 1 October 2014).

Priest, J. (2004) *The State of Wireless London*. http://informal.org.uk/people/julian/publications/the_state_of_wireless_london/ (Accessed: 1 October 2014).

Proshansky, H., Ittelson, W. and Rivlin, L. (eds) (1970) *Environmental Psychology: Man and His Physical Setting*. New York: Holt, Rinehart & Winston.

Putnam, R.D. (2000) *Bowling Alone: The Collapse and Revival of American Community*. New York and London: Simon & Schuster.

Rapoport, A. (1981) 'Identity and Environment: a Cross-cultural Perspective', in Duncan, J.S. (ed.), *Housing and Identity: Cross-cultural Perspectives*. London: Croom Helm.

Rapoport, A. (1982) *The Meaning of the Built Environment: A Nonverbal Communication Approach*. London: Sage Publications.

Rapoport, A. (1994) 'Spatial organization and the built environment', in Ingold, T. (ed.), *Companion Encyclopedia of Anthropology: Humanity, Culture and Social Life*. Abingdon: Routledge, pp. 460–502.

Rayner, R. (1998) 'Nowhere, U.S.A.' *The New York Times*. http://www.nytimes.com/1998/03/08/magazine/nowhere-usa.html, 8 March 1998.

Relph, E. (1984) *Place and Placelessness*. London: Pion Limited.

Rheingold, H. (2000) *The Virtual Community: Homesteading on the Electronic Frontier* Cambridge, MA: The MIT Press.

Rheingold, H. (2003) *Smart Mobs: The Next Social Revolution*. New York: Basic Books.

Riley, T. (1999) 'The Un-Private House'. *The Un-Private House*. New York: The Museum of Modern Arts, New York.

Robinson, D. and Davies, S. (2014) 'Online sales lift small takeways up the pecking order'. *Financial Times*. http://www.ft.com/cms/s/0/dacdea8c-bb3a-11e3-b2b7-00144feabdc0.html-axzz3NwaVZUzL, 3 April 2014.

Rose, D. (2014) *Enchanted Objects: Design, Human Desire, and the Internet of Things*. New York: Scribner Book Company.

Rosen, J. (2010) 'The web means the end of forgetting'. http://www.nytimes.com/2010/07/25/magazine/25privacy-t2.html?pagewanted=all&_r=0, 21 July 2010.

Sassen, S. (2006a) *Territory, Authority, Rights: From Medieval to Global Assemblages*. Princeton, NJ: Princeton University Press.

Sassen, S. (2006b) 'Making Public Interventions in Today's Massive Cities'. *Static*. http://static.londonconsortium.com/issue04/sassen_publicintervensions.php (Accessed: 2014).

Sassen, S. (2011a) 'Talking back to your intelligent city'. *Voices on Society*; *McKinsey and Company*. http://voices.mckinseyonsociety.com/talking-back-to-your-intelligent-city/ (Accessed: 1 October 2014).

Sassen, S. (2012) 'Urbanising technology', in Burdett, R. and Rode, P. *The Electric City Newspaper*. http://ec2012.lsecities.net/newspaper/: LSE Cities. 12–14.

Sassen, S. (2011b) 'An interview with Saskia Sassen about "Smart cities"'. http://www.nicolasnova.net/pasta-and-vinegar/2011/07/06/an-interview-with-saskia-sassen-about-smart-cities (Accessed: 13 August 2014).

Scannell, P. (1996) *Radio, Television and Modern Life: A Phenomenological Approach*. Oxford: Blackwell.

Schegloff, E. (2002) 'Beginnings in the telephone', in Katz, J.E. and Aakhus, M. (eds), *Perpetual Contact: Mobile Communication, Private Talk, Public Performance*. Cambridge: Cambridge University Press, pp. 284–300.

gen. Schieck, A.F., Schnädelbach, H., Motta, W., Behrens, M., North, S., Ye, L. and Kostopoulou, E. (2014) 'Screens in the Wild: Exploring the Potential of Networked Urban Screens for Communities and Culture', in Gehring, S. (ed.), *The International Symposium on Pervasive Displays (PerDis '14)*. Copenhagen, Denmark June 3rd–June 4th 2014. ACM, pp. 166–167.

Schivelsbusch, W. (1995) *Disenchanted Night*. Berkeley, CA: University of California Press.

Schmidt, M. and Schmitt, E. (2013) 'Obama's Portable Zone of Secrecy (Some Assembly Required)'. *The New York Times*, 9 November 2013.

Schofield, H. (2014) 'Paris car ban imposed after pollution hits high'. *BBC*. http://www.bbc.com/news/world-europe-26599010, 18 March 2014.

Schwartz, R. (2014) 'Online Place Attachment: Exploring Technological Ties to Physical Places', in de Souza e Silva, A. and Sheller, M. (eds), *Mobility and Locative Media: Mobile Communication in Hybrid Spaces*. Abingdon: Routledge, pp. 85–100.

Scoble, R. and Israel, S. (2014) *Age of Context: Mobile, Sensors, Data and the Future of Privacy*. USA: Patrick Brewster Press.

Sennett, R. (1977) *The Fall of Public Man: On the Social Psychology of Capitalism*. New York: Alfred A. Knopf.

Sennett, R. (2010) 'The Public Realm', in Bridge, G. and Watson, S. (eds), *The Blackwell City Reader*. Chichester, UK: Wiley-Blackwell, pp. 261–272.

Sheller, M. and Urry, J. (2000) 'The city and the car'. *Int. J. Urban Reg. Res.*, 24 (4), pp. 737–757.

Sheller, M. and Urry, J. (2006) 'Introduction: Mobile Cities, Urban Mobilities', in Sheller, M. and Urry, J. (eds), *Mobile Technologies of the City*. Oxford: Routledge, pp. 1–17.

Shepard, M. (2010) 'On Hertzian Space and Urban Architecture'. *Vague Terrain*, 16.

Shepard, M. (ed.) (2011) *Sentient City: Ubiquitous Computing, Architecture, and the Future of Urban Space*. Cambridge, MA: The MIT Press.

Simmel, G. (1950) *The Sociology of Georg Simmel*, trans., ed. and introduction by Wolff, K.H., Glencoe, IL: Free Press.

Simmel, G. (1971) *On Individuality and Social Forms: Selected Writings [of] Georg Simmel*, ed. Levine, D.N. Chicago and London: University of Chicago Press.

Simmel, G. (1994 [1909]) 'Bridge and Door'. *Theory, Culture & Society*, 11, pp. 5–10.

Simonite, T. (2012) 'What Facebook Knows'. *MIT Technology Review*.

Solow, R. (1987) 'We'd better watch out'. *The New York Times Book Review*. *The New York Times*.

Solsman, J.E. (2013) 'Netflix, YouTube gobble up half of Internet traffic'. *CNet*. http://www.cnet.com/uk/news/netflix-youtube-gobble-up-half-of-internet-traffic/ (Accessed: 1 October 2014).

Southern, J. and Speed, C. (2009) 'CoMob'. *International Symposium for Electronic Art*. Manchester: ISEA, p. 72.

Southern, J. and Speed, C. (2013) *Co-Mob Research*. http://www.comob.org.uk/?page_id=179 (Accessed: 1 October 2014).

Star, S.L. (1999) 'An Ethnography of Infrastructure'. *American Behavioral Scientist*, 43 (3), pp. 377–391.

Star, S.L. and Ruhleder, K. (1996) 'Steps Toward an Ecology of Infrastructure: Design and Access for Large Information Spaces'. *Information Systems Research*, 7 (1), pp. 111–134.

Steiner, R. (2013) 'The house of the future: London property is kitted out with iPads in the walls, remote-controlled coffee pots, magnetic wallpaper and a 4 METRE widescreen TV'. *The Daily Mail*. http://www.dailymail.co.uk/sciencetech/article-2360496/London-property-kitted-iPads-walls-remote-controlled-coffee-pots-magnetic-wallpaper-4-METRE-widescreen-TV.html, 11 July 2013.

Stevens, Q. (2007) 'Betwixt and Between: Building Thresholds, Liminality and Public Space', in Franck, K. and Stevens, Q. (eds), *Loose Space: Possibility and Diversity in Urban Life*. Abingdon and New York: Taylor & Francis, pp. 73–92.

Struppek, M. (2006) 'The social potential of Urban Screens'. *Visual Communication*, 5 (2), pp. 173–188.

Struppek, M. (2014) 'Urban Media Cultures Reflecting Modern City Development'. *Screen City Journal*, May 2014.

Suchman, L.A. (1987) *Plans and Situated Actions: The Problem of Human-Machine Communication*. Cambridge: Cambridge University Press.

Sullivan, M. (2011) '20 Best U.S. Airports for Tech Travelers'. *Computer World*. http://www.computerworld.com/article/2500425/enterprise-applications/20-best-u-s--airports-for-tech-travelers.html?page=2 (Accessed: 1 October 2014).

Sung, J.-Y., Guo, L., Grinter, R.E. and Christensen, H.I. (2007) '"My Roomba is Rambo": intimate home appliances'. *Proceedings of the 9th International Conference on Ubiquitous Computing*. Innsbruck, Austria: Springer-Verlag, pp. 145–162.

Sutherland, I.E. (1965) 'The Ultimate Display', *Congress of the Internation Federation of Information Processing IFIP*. New York City, 24–29 May 1965, pp. 506–508.

Swyngedouw, E. (1993) 'Communication, Mobility and the Struggle for Power over Space', in Giannopoulos, G. and Gillespie, A. (eds), *Transport and Communications Innovation in Europe*. London: Belhaven Press, pp. 305–325.

Taipei Government *Taipei Free Wifi*. http://www.tpe-free.taipei.gov.tw/tpe/index_en.aspx (Accessed: 1 October 2014).

Tallentyre, M. (2013) 'Failed sat nav blamed for Chester-le-Street cyclist's A68 death'. *Northern Echo*.

TaskRabbit *How Does TaskRabbit Work?* https://www.taskrabbit.com/how-it-works (Accessed: 1 January 2015).

Taylor, K. (2006) 'Programming video art for urban screens in public space', *First Monday* Special Issue #4: Urban Screens: Discovering the potential of outdoor screens for urban society.

Thackara, J. (2006) *In the Bubble: Designing in a Complex World*. Cambridge, MA and London: The MIT Press.

The Metropolitan Museum of Art (n.n.) *Leap into the void*. http://www.metmuseum.org/toah/works-of-art/1992.5112 (Accessed: 24 August 2014).

Thompson, C. (2013) 'How to Keep the NSA Out of Your Computer: Sick of government spying, corporate monitoring, and overpriced ISPs? There's a cure for that'. *Mother Jones*. http://www.motherjones.com/politics/2013/08/mesh-internet-privacy-nsa-isp (Accessed: 1 October 2014).

Thomson, I. (2014) 'Facebook goes down, people dial 911. Police appeal for calm – yes, seriously'. *The Register*. http://www.theregister.co.uk/2014/08/01/facebook_outage/ (Accessed: 1 August 2014).

The Economist (2013). *The rise of the sharing economy*. http://www.economist.com/news/leaders/21573104-internet-everything-hire-rise-sharing-economy (Accessed: 1 August 2014).

Thrift, N. and French, S. (2002) 'The Automatic Production of Space'. *Transactions of the Institute of British Geographers*, 27 (3), pp. 309–335.

Thrift, N. and May, J. (2001) *Timespace: Geographies of Temporality*. Routledge Critical Geographies Series. New York: Routledge.

Thuma, A. (2011) 'Hannah Arendt, Agency, and the Public Space', Behrensen, M., Lee, L. and Tekelioglu, A.S. (eds), *Modernities Revisited*. Vienna IWM Junior Visiting Fellows Conferences.

Townsend, A. (2000) 'Life in the real-time city: mobile telephones and urban metabolism'. *Journal of Urban Technology*, 7 (2), pp. 85–104.

Townsend, A. (2003) *Wired/Unwired: The Urban Geography of Digital Networks*. Massachusetts Institute of Technology.

Tuan, Y.-F. (1977) *Space and Place: The Perspective of Experience*. Minneapolis: University of Minnesota Press.

Tufekci, Z. and Wilson, C. (2012) 'Social Media and the Decision to Participate in Political Protest: Observations From Tahrir Square'. *Journal of Communication*, 62 (2), pp. 363–379.

Turk, V. (2013) How to set up a smart house. *Wired.co.uk*.

Turkle, S. (2011) *Alone Together: Why We Expect More from Technology and Less from Each Other*. New York: Basic Books.

Tuters, M. (2004) 'The Locative Commons: Situating Location-Based Media in Urban Public Space'. *Futuresonic 2004*. Manchester, UK.

Twilio. https://www.twilio.com/elements (Accessed: 1 October 2014).

UK Online Centres (2008) *Digital inclusion, social impact: a research study*. http://www.tinderfoundation.org/sites/default/files/research-publications/digital_inclusion_research_report.pdf: Tinder Foundation.

Undergleider, N. (2012) 'Bloomberg On Mayors Vs. Foursquare Mayors'. *Fast Company*. http://www.fastcompany.com/1826520/bloomberg-mayors-vs-foursquare-mayors (Accessed: 1 October 2014).

Urry, J. (2003) 'Social Networks, Travel and Talk'. *British Journal of Sociology*, 54 (2), pp. 155–175.

US Air Force (2014) *Timing*. http://www.gps.gov/applications/timing/ (Accessed: 1 October 2014).

US Department of Energy (2012) *International Energy Statistics*. 2014 (2 October 2014) http://www.eia.gov/cfapps/ipdbproject/iedindex3.cfm?tid=2&pid=2&aid=2&cid=r3,&s yid=2008&eyid=2012&unit=BKWH: US Department of Energy.

Vanderbilt, O. (2009) 'Data Center Overload'. *The New York Times*. http://www.nytimes. com/2009/06/14/magazine/14search-t.html?pagewanted=all&_r=0, 8 June 2009.

Varnelis, K. (ed.) (2012) *Networked Publics*. Cambridge, MA: The MIT Press.

Veolia Environment (2013) *Imagine 2050 – Veolia*. http://www.veolia.co.uk/media/research (Accessed: 1 October 2014).

Visitberlin.de *W-LAN for all: "Public Wi-Fi Berlin": New Hotspots in the Capital*. Berlin Tourismus and Kongress GmbH. http://www.visitberlin.de/en/article/w-lan-for-all-public-wi-fi-berlin (Accessed: 1 October 2014).

van de Vliet, V. (2013) 'Space for People, Not for Cars'. *Works That Work 1*. https://worksthatwork.com/1/shared-space (Accessed: 1 October 2014).

de Waal, M. (2011) 'The Urban Culture of Sentient Cities: From an internet of things to a public sphere centered around things', in Shepard, M. (ed.), *Ubiquitous Computing, Architecture, and the Future of Urban Space*. Cambridge, MA: The MIT Press.

Ward, M. (2002) 'How mobile phone masts "vanish"'. *BBC News*. Monday, 16 September 2002. http://news.bbc.co.uk/1/hi/in_depth/sci_tech/2000/dot_life/2261039.stm (Accessed: 1 October 2014).

Ward, M. (2014) 'Browsing speeds may slow as net hardware bug bites'. *BBC News*. 14 August 2014. http://www.bbc.co.uk/news/technology-28786954 (Accessed: 1 October 2014).

Warschauer, M. (2003) *Technology and Social Inclusion: Rethinking the Digital Divide*. Cambridge, MA: The MIT Press.

Wasik, B. (2010) *And Then There's This: How Stories Live and Die in Viral Culture*. New York: Penguin.

Weiser, M. (1991) 'The Computer for the 21st Century'. *Scientific American*, 265 (3), pp. 94–104.

Wellesley-Miller, S. (1976) 'Intelligent Environments', in Negorponte, N. (ed.), *Soft Architecture Machines*. Cambridge, MA: The MIT Press.

Wellman, B. (ed.) (1999) *Networks in the Global Village: Life in Contemporary Communities*. Boulder, CO: Westview Press.

Wellman, B. (2001a) 'Little Boxes, Glocalization, and Networked Individualism', in Tanabe, M., van den Besselaar, P. and Ishida, T. (eds), *Second Kyoto Workshop on Digital Cities II, Computational and Sociological Approaches*. London: Springer-Verlag, pp. 10–25.

Wellman, B. (2001b) 'Physical Place and Cyberplace: The Rise of Personalized Networking'. *International Journal of Urban and Regional Research*, 25 (2), pp. 227–252.

Wellman, B., Haase, A.Q., Witte, J. and Hampton, K. (2001) 'Does the Internet Increase, Decrease, or Supplement Social Capital?: Social Networks, Participation, and Community Commitment'. *American Behavioral Scientist*, 45 (3), pp. 436–455.

Westin, A.F. (1967) *Privacy and Freedom*. New York: Atheneum.

Whyte, W. (1988) *City: Rediscovering the Center*. New York: Doubleday.

Whyte, W.H. (1980) *The Social Life of Small Urban Space*. Washington, DC: The Conservation Foundation.

Wifi in Estonia. (2014). http://www.visitestonia.com/en/things-to-know-about-estonia/facts-about-estonia/wifi-in-estonia (Accessed: 1 October 2014).

Wiig, A. (2013) 'Everyday Landmarks of Networked Urbanism: Cellular Antenna Sites and the Infrastructure of Mobile Communication in Philadelphia'. *Journal of Urban Technology*, 20 (3), pp. 21–37.

Wikipedia (n.n.) *Criticism of Facebook*. Wikipedia. http://en.wikipedia.org/wiki/Criticism_of_Facebook (Accessed: 1 October 2014).

Wikipedia *Telecentre*. Wikipedia. http://en.wikipedia.org/wiki/Telecentre (Accessed: 1 October 2014).

Wikle, T. (2002) 'Cellular Tower Proliferation in the United States'. *Geographical Review*, 92 (1), pp. 45–62.

Williams, R. (1974) *Television, Technology and Cultural Form*. London: Fontana.

Willis, K., Roussos, G., Chorianopoulos, K. and Struppek, M. (2008) *Shared Encounters*. London and New York: Springer.

Willis, K.S. (2006) *Wayfinding Situations*. Bauhaus University Weimar.

Willis, K.S. (2008) 'Places, Situations and Connections', in Aurigi, A. and Cindio, F.D. (eds), *Augmented Urban Spaces: Articulating the Physical and Electronic City*. Aldershot: Ashgate, pp. 9–26.

Willis, K.S. (2012) 'Being in Two Places at Once', in Abend, P., Haupts, T. and Müller, C. (eds), *Medialität der Nähe: Situationen – Praktiken – Diskurse*. Bielefeld: Transcript.

Winston, B. (1998) *Media Technology and Society. A History From the Telegraph to the Internet*. London: Routledge.

Wireless België. (2014). http://www.wirelessantwerpen.be/ (Accessed: 1 October 2014).

Yue, A. (2009) 'Urban Screens, spatial regeneration and cultural citizenship the embodied interaction of cultural participation', in McQuire, S., Martin, M. and Niederer, S. (eds), *The Urban Screens Reader*. Amsterdam: Institute of Network Cultures, pp. 261–278.

Zheng, Y. (2011) 'Location-Based Social Networks: Users', in Zheng, Y. and Zhou, X. (eds), *Computing with Spatial Trajectories*. New York: Springer.

Zickuhr, K. and Smith, A. (2012) *Digital differences*. http://www.pewinternet.org/2012/04/13/digital-differences/ – fn 188–20: Pew Research Internet Project.

Zuckerberg, M. (2014) Facebook. https://www.facebook.com/zuck/posts/10101322049893211 (Accessed: 13 August 2014).

Index